CHANCE
and
NECESSITY

CHANCE
AND
NECESSITY

An Essay on the Natural
Philosophy of Modern Biology

by JACQUES MONOD

translated from the French by Austryn Wainhouse

ALFRED A. KNOPF
New York 1971

THIS IS A BORZOI BOOK
PUBLISHED BY ALFRED A. KNOPF, INC.

First American Edition

ISBN: 0-394-46615-2

Library of Congress Catalog Card Number: 77-154929

Manufactured in the United States of America

At that subtle moment when man glances backward over his life, Sisyphus returning toward his rock, in that slight pivoting he contemplates that series of unrelated actions which becomes his fate, created by him, combined under his memory's eye and soon sealed by his death. Thus, convinced of the wholly human origin of all that is human, a blind man eager to see who knows that the night has no end, he is still on the go. The rock is still rolling.

I leave Sisyphus at the foot of the mountain! One always finds one's burden again. But Sisyphus teaches the higher fidelity that negates the gods and raises rocks. He too concludes that all is well. This universe henceforth without a master seems to him neither sterile nor futile. Each atom of that stone, each mineral flake of that night-filled mountain, in itself forms a world. The struggle itself toward the heights is enough to fill a man's heart. One must imagine Sisyphus happy.

— ALBERT CAMUS, *The Myth of Sisyphus*

Contents

Contents

Contents

Preface

BIOLOGY OCCUPIES a position among the sciences at once marginal and central. Marginal because – the living world constituting but a tiny and very "special" part of the universe – it does not seem likely that the study of living beings will ever uncover general laws applicable outside the biosphere. But if the ultimate aim of the whole of science is indeed, as I believe, to clarify man's relationship to the universe, then biology must be accorded a central position, since of all the disciplines it is the one that endeavors to go most directly to the heart of the problems that must be resolved before that of "human nature" can even be framed in other than metaphysical terms.

Consequently no other science has quite the same significance for man; none has already so heavily contributed to the shaping of modern thought, profoundly and definitively affected as it has been in every domain – philosophical, religious, political – by the advent of the theory of evolution. Its phenomenological validity generally accepted by the close of the last century, the theory of evolution, while dominating the whole of biology, yet remained as if suspended, awaiting the elaboration of a *physical* theory of heredity. Thirty years ago, the hope that one would soon be forthcoming appeared almost illusory, notwithstanding the successes in classical genetics. Today, however, this is precisely what we have in the molecular theory of the genetic code. Here "theory of the genetic code" is to be understood in the broader sense, including not only concepts relevant to the chemical structure of hereditary material and the information it conveys, but also the molecular mechanisms for expressing this information morphogenetically and physiologically. So defined, the theory of the genetic code constitutes the fundamental basis of biology. This does not mean, of course, that the complex structures and functions of organisms can be *deduced* from it, nor even that they are always directly analyzable on the molecular level. (Nor can everything in chemistry be predicted or resolved by means of the quantum theory, which, beyond any question, underlies all chemistry.)

But although the molecular theory of the code cannot now—and will doubtless never be able to—predict and resolve the whole of the biosphere, it does today constitute a general theory of living systems. No such thing existed in scientific knowledge prior to molecular biology. Until then the "secret of life" could be viewed as essentially inaccessible. In recent times much of it has been laid bare. This, a considerable event, ought certainly to make itself strongly felt in contemporary thinking, once the general significance and consequences of the theory are understood and appreciated beyond the narrow circle of specialists. I hope the present essay will be useful to that end. In it, rather than making a thorough survey of the contents of modern thinking in biology, I have tried to bring out the "form" of its key concepts and to point to their logical relationships with other areas of thought.

Preface

Nowadays a man of science is not well advised to use the word *philosophy*, albeit qualified as *natural*, in the title (or even subtitle) of a book: nothing more is needed to earn it a distrustful reception from other scientists, and from philosophers a condescending one at best. I have only one excuse, but I believe it is sound: the duty which more forcibly than ever thrusts itself upon scientists to apprehend their discipline within the larger framework of modern culture, with a view to enriching the latter not only with technically important findings, but also with what they may feel to be humanly significant ideas arising from their area of special concern. The very ingenuousness of a fresh look at things (and science possesses an ever-youthful eye) may sometimes shed a new light upon old problems.

Meanwhile, to be sure, any confusion between the ideas *suggested* by science and science itself must be carefully avoided; but it is just as necessary that scientifically warranted conclusions be resolutely pursued to the point where their full meaning becomes clear. A difficult exercise. I do not claim to have turned in a faultless performance. First let me say that in what follows the strictly biological part is in no sense original: I have done no more than summarize what are considered established ideas in contemporary science. The relative weighting of various developments, like the choice of examples offered, does, it is true, reflect personal tendencies. Some outstanding chapters in biology are not even mentioned. But—once again—this essay does not seek to discuss the entire field of biology; it is an avowed attempt to extract the quintessence of the molecular theory of the code. For the ideological generalizations I have ventured to deduce from it I am, of course, solely responsible. But I do not think I am mistaken in saying that, where they do not exceed the bounds of epistemology, these interpretations would find assent from the majority of modern biologists. I must claim full responsibility as well for the ethical and sometimes political ideas I have expressed and preferred not to avoid, perilous though they are and however naïve or overambitious they may appear. Modesty befits the scientist, but not the ideas that inhabit him and which he is under the obligation of upholding. There again, however, I have the strengthening assurance of finding myself in full agreement with certain contemporary biologists whose achievements are worthy of the highest regard.

I must beg the indulgence of biologists for pages of what will strike them as tediously self-evident explanation; of nonbiologists for the dryness of other pages given over to unavoidable "technical" background. Some readers may be helped over these difficulties by the appendixes. But I should like to stress that they can be dispensed with by anyone who is not disposed to grapple directly with the chemical realities of biology.

This essay grows out of a series of lectures – the Robbins Lectures – given in February of 1969 at Pomona College in California. I wish to thank the authorities of that college for having provided me with the occasion to explore, before a very young and eager audience, certain themes which had for a long time been with me a subject for thought, but not for teaching. These themes were also at the core of a course I gave at the Collège de France during the 1969–70 academic year. A fine and precious institution it is that allows its members sometimes to step beyond the strict boundaries of their charge and purview. Thanks therefor be unto Guillaume Budé and King Francis I.

CHANCE
AND
NECESSITY

I
Of Strange Objects

THE DIFFERENCE between artificial and natural objects seems immediately and unambiguously apparent to all of us. A rock, a mountain, a river, or a cloud – these are natural objects; a knife, a handkerchief, a car – so many artificial objects, artifacts.* Analyze these judgments, however, and it will be seen that they are neither immediate nor strictly objective. We know that the knife was man-made for a use its maker visualized beforehand. The object renders in material form the preexistent intention that gave birth to it, and its form is accounted for by the performance expected of it even before it takes shape. It is another story altogether with the river or the rock which we know, or believe, to have been molded by the free play of physical forces to which we cannot attribute any design, any "project" or purpose. Not, that is, if we accept the basic premise of the scientific method, to wit, that nature is *objective* and not *projective*.

Hence it is through reference to our own activity, conscious and projective, intentional and purposive – it is as

---

*In the literal sense: products of human art or workmanship.

makers of artifacts—that we judge of a given object's "naturalness" or "artificialness." Might there be objective and general standards for defining the characteristics of artificial objects, products of a conscious purposive activity, as against natural objects, resulting from the gratuitous play of physical forces? To make sure of the complete objectivity of the criteria chosen, it would doubtless be best to ask oneself whether, in putting them to use, a program could be drawn up enabling a computer to distinguish an artifact from a natural object.

Such a program could be applied in the most interesting connections. Let us suppose that a spacecraft is soon to be landed upon Venus or Mars; what more fascinating question than to find out whether our neighboring planets are, or at some earlier period have been, inhabited by intelligent beings capable of projective activity? In order to detect such present or past activity we would have to search for and be able to recognize its *products*, however radically unlike the fruit of human industry they might be. Wholly ignorant of the nature of such beings and of the projects they might have conceived, our program would have to utilize only very general criteria, solely based upon the examined objects' structure and form and without any reference to their eventual function.

The suitable criteria, we see, would be two in number: (a) regularity, and (b) repetition. By means of the first one would seek to make use of the fact that natural objects, wrought by the play of physical forces, almost never present geometrically simple and straightforward structures: flat surfaces, for instance, or rectilinear edges, right angles, exact symmetries; whereas artifacts will ordinarily show such features, if only in an approximate or rudimentary manner.

Of the two criteria, repetition would probably be the more decisive. Materializing a reiterated intent, homologous artifacts meant for the same use reflect, faithfully in the main, the constant purpose of their creator. In that respect the discovery of numerous specimens of closely similar objects would be of high significance.

These, briefly defined, are the general criteria that might serve. The objects selected for examination, it must be added, would be of *macroscopic* dimensions, but not *microscopic*. By macroscopic is meant dimensions measurable, say, in centimeters; by microscopic, dimensions normally expressed in angstroms (a hundred million of which equal one centimeter). This proviso is crucial, for on the microscopic scale one would be dealing with atomic and molecular structures whose simple and repetitive geometries, obviously, would attest not to a conscious and rational intention but to the laws of chemistry.

Now let us suppose the program drawn up and the machine built. To check its performance, the best possible test would be to put it to work upon terrestrial objects.

difficulties of a space program

Let us invert our hypotheses and imagine that the machine has been put together by the experts of a Martian NASA aiming at detecting evidence of organized, artifact-producing activity on Earth. And let us suppose that the first Martian craft comes down in the Forest of Fontainebleau, not far, let's say, from the village of Barbizon. The machine looks at and compares the two series of objects most prominent in the area: on the one hand the houses in Barbizon, on the other hand the rock formations of Apremont. Utilizing the criteria of regularity, of geometric simplicity, and of repetition, it will have no trouble deciding that the rocks are natural objects and the houses artifacts.

Focusing now upon lesser objects, the machine examines some pebbles, near which it discovers some crystals—quartz crystals, let us say. According to the same criteria it should of course decide that while the pebbles are natural, the quartz crystals are artificial objects. A decision which appears to point to some "error" in the writing of the program. An "error" which, moreover, proceeds from an interesting source: if the crystals present perfectly defined geometrical shapes, that is because their macroscopic structure directly reflects the simple and repetitive microscopic structure of the atoms or molecules constituting them. A crystal, in other words, is the macroscopic expression of a microscopic structure. An "error" which, by the by, should be easy enough to eliminate, since all *possible* crystalline structures are known to us.

But let us suppose that the machine is now studying another kind of object: a hive built by wild bees, for example. There it would obviously find all the signs indicating artificial origin: the simple and repeated geometrical structures of the honeycombs and the cells composing them, thanks to which the hive would earn classification in the same category of objects as the Barbizon dwellings. What are we to make of this conclusion? We know the hive is "artificial" insofar as it represents the product of the activity of bees. But we have good reasons for thinking that this activity is strictly automatic—immediate, but not consciously projective. At the same time, as good naturalists we view bees as "natural" beings. Is there not a flagrant contradiction in considering "artificial" the

6

product of a "natural" being's automatic activity?

Carrying the investigation a little further, it would soon be seen that if there is contradiction, it results not from faulty programming but from the ambiguity of our judgments. For if the machine now inspects, not the hive, but the bees themselves, it cannot take them for anything but artificial, highly elaborated objects. The most superficial examination will reveal in the bee elements of simple symmetry: bilateral and translational. Moreover and above all, examining bee after bee the computer will note that the extreme complexity of their structure (the number and position of abdominal hairs, for example, or the ribbing of the wings) is reproduced with extraordinary fidelity from one individual bee to the next. Powerful evidence, is it not, that these creatures are the products of a deliberate, constructive, and highly sophisticated order of activity? Upon the basis of such conclusive documentation, the machine would be bound to signal to the officials of the Martian NASA its discovery, upon Earth, of an industry compared with which their own would probably seem primitive.

In this little excursion into the not-so-very-farfetched, our aim was only to illustrate the difficulty of defining the distinction—elusive, for all its obviousness to our intuitions—between "natural" and "artificial" objects. In fact, on the basis of structural criteria, macroscopic ones, it is probably impossible to arrive at a definition of the artificial which, while including all "veritable" artifacts, such as the products of human workmanship, would exclude objects so clearly natural as crystalline structures, and indeed, the living beings themselves which we would also like to classify among natural systems.

Looking for the cause of the confusion – or in any case, seeming confusion – the program is leading to, we may perhaps wonder whether it does not arise from our having wished to limit it to considerations only of form, of structure, of geometry, and so divesting our notion of an artificial object of its essential content. This being that any such object is defined or explained primarily by the function it is intended to fulfill, the performance its inventor expects of it. However, we shall soon find that by programming the machine so that henceforth it studies not only the structure but the eventual performance of the examined objects, we end up with still more disappointing results.

For let us suppose that this new program does enable the machine to analyze correctly the structure and the performance of two series of objects – horses running in a field and automobiles moving on a highway, for example. The analysis would tend to the conclusion that these objects are closely comparable, those making up each series having a built-in capacity for swift movement, although over different surfaces, which accounts for their differences of structure. And if, to take another example, we were to ask the machine to compare the structure and performance of the eye of a vertebrate with that of a camera, the program would have to acknowledge their profound similarities: lenses, diaphragm, shutter, light-sensitive pigments: surely, the same components could not have been introduced into both objects except with a view to getting similar performances from them.

The last of these examples is a classic one of functional

objects endowed with a purpose

adaptation in living beings, and I have cited it only to emphasize how arbitrary and pointless it would be to deny that the natural organ, the eye, represents the materialization of a "purpose" – that of picking up images – while this is indisputably also the origin of the camera. It would be the more absurd to deny it since, in the last analysis, the purpose which "explains" the camera can only be the same as the one to which the eye owes its structure. Every artifact is a product made by a living being which through it expresses, in a particularly conspicuous manner, one of the fundamental characteristics common to all living beings without exception: that of being *objects endowed with a purpose or project*, which at the same time they exhibit in their structure and carry out through their performances (such as, for instance, the making of artifacts).

Rather than reject this idea (as certain biologists have tried to do) it is indispensable to recognize that it is essential to the very definition of living beings. We shall maintain that the latter are distinct from all other structures or systems present in the universe through this characteristic property, which we shall call *teleonomy*.

But it must be borne in mind that, while necessary to the definition of living beings, this condition is not sufficient, since it does not propose any objective criteria for distinguishing between living beings themselves and the artifacts issuing from their activity.

It is not enough to point out that the project which gives rise to an artifact belongs to the animal that created it, and not to the artificial object itself. This obvious notion is also too subjective, as the difficulty of utilizing it in the computer program would prove: for upon what basis would the machine be able to decide that the project of picking up images – the project represented by the camera – belongs to some object other than the camera itself? By examining nothing beyond the finished structure and by simply analyzing its performance it is possible to identify the project, but not its author or source.

To achieve this we must have a program which studies not only the actual object but its origin, its history, and, for a start, how it has been put together. Nothing, in principle at least, stands in the way of formulating such a program. Even if it were rather crudely compiled, we would be able with it to discern a radical difference between any artifact, however highly perfected, and a living being. The machine could not fail to note that the macroscopic structure of an artifact (whether a honeycomb, a dam built by beavers, a paleolithic hatchet, or a spacecraft) results from the application to the materials constituting it of forces *exterior* to the object itself. Once complete, this macroscopic structure attests, not to inner forces of cohesion between atoms or molecules constituting its material (and conferring upon it only its general properties of density, hardness, ductility, etc.), but to the *external* forces that have shaped it.

On the other hand, the program will have to register the fact that a living being's structure results from a totally different process, in that it owes almost nothing to the action of outside forces, but everything, from its overall shape down to its tiniest detail, to "morphogenetic" interactions within the object itself. It is thus a structure giving proof of an autonomous determinism: precise, rigorous, implying a

self-constructing machines

10

virtually total "freedom" with respect to outside agents or conditions – which are capable, to be sure, of impeding this development, but not of governing or guiding it, not of prescribing its organizational scheme to the living object. Through the autonomous and spontaneous character of the morphogenetic processes that build the macroscopic structure of living beings, the latter are absolutely distinct from artifacts, as they are, furthermore, from the majority of natural objects whose macroscopic morphology largely results from the influence of external agents. To this there is a single exception: that, once again, of crystals, whose characteristic geometry reflects microscopic interactions occurring within the object itself. Hence, utilizing this criterion alone, crystals would have to be classified together with living beings, while artifacts and natural objects, alike fashioned by outside agents, would comprise another class.

That this last criterion, after those of regularity and repetition, should point to a similarity between crystalline structures and the structures of living beings might well set our programmer to thinking. Though unversed in modern biology, he would be obliged to wonder whether the internal forces which confer their macroscopic structure upon living beings might be of the same nature as the microscopic interactions responsible for crystalline morphologies. That this is indeed the case constitutes one of the main themes to be developed in the ensuing chapters of this essay. But for the moment we are looking for the most general criteria to define the macroscopic properties that set living beings apart from all other objects in the universe.

Having "discovered" that an internal, autonomous determinism guarantees the formation of the extremely complex structures of living beings, our programmer (with no training in biology, but an information specialist by profession) must necessarily see that such structures represent a considerable quantity of information whose source has still to be identified: for all expressed – and hence received – information presupposes a source.

Let us assume that, continuing his investigation, our programmer at last makes his final discovery: that the source of the information expressed in the structure of a living being is *always* another, structurally identical object. He has now identified the source and detected a third remarkable property in these objects: their ability to reproduce and to transmit *ne varietur* the information corresponding to their own structure. A very rich body of information, since it describes an organizational scheme which, along with being exceedingly complex, is preserved intact from one generation to the next. The term we shall use to designate this property is *invariant reproduction*, or simply *invariance*.

self-reproducing machines

With their invariant reproduction we find living beings and crystalline structures once again sharing a property that renders them unlike all other known objects in the universe. Certain chemicals in supersaturated solution do not crystallize unless the solution has been inoculated with crystal seeds. We know as well that in cases of a chemical capable of crystallizing into two different systems, the structure of the crystals appearing in the solution will be determined by that of the seed employed. Crystalline structures, however, represent a quantity of information by several orders of magnitude inferior to that transmitted from one generation to another in the simplest living beings we are acquainted with. By this criterion—purely quantitative, be it noted—living beings may be distinguished from all other objects, crystals included.

Let us now forget our Martian programmer and leave him to mull things over undisturbed. This imaginary experiment has had no other aim than to compel us to "rediscover" the more general properties that characterize living beings and distinguish them from the rest of the universe. Let us now admit to a familiarity with modern biology, so as to go on to analyze more closely and to try to define more precisely, if possible quantitatively, the properties in question. We have found three: teleonomy, autonomous morphogenesis, and reproductive invariance.

Of them all, reproductive invariance is the least difficult to define quantitatively. Since this is the capacity to reproduce highly ordered structure, and since a structure's degree of order can be defined in units of information, we shall say that the "invariance content" of a given species is equal to the amount of information which, transmitted from one generation to the next, assures the preservation of the specific structural standard. As we shall see later on, with the help of a few assumptions it will be possible to arrive at an estimate of this amount.

**strange properties:
invariance
and teleonomy**

That in turn will enable us to bring into better focus the notion most immediately and plainly inspired by the examination of the structures and performances of living beings, that of teleonomy. Analysis nevertheless reveals it to be a profoundly ambiguous concept, since it implies the subjective idea of "project." We remember the example of the camera: if we agree that this object's existence and structure realize the "project" of capturing images, we must also agree, obviously enough, that a similar project is accomplished with the emergence of the eye of a vertebrate.

But it is only as a part of a more comprehensive project that each individual project, whatever it may be, has any meaning. All the functional adaptations in living beings, like all the artifacts they produce, fulfill particular projects which may be seen as so many aspects or fragments of a unique primary project, which is the preservation and multiplication of the species.

To be more precise, we shall arbitrarily choose to define the essential teleonomic project as consisting in the transmission from generation to generation of the invariance content characteristic of the species. All the structures, all the performances, all the activities contributing to the success of the essential project will hence be called "teleonomic."

This allows us to put forward at least the *principle* of a definition of a species' "teleonomic level." All teleonomic structures and performances can be regarded as corresponding to a certain quantity of information which must be transmitted for these structures to be realized and these performances accomplished. Let us call this quantity "teleonomic information." A given species' "teleonomic level" may then be said to correspond to the quantity of information which, on the average and per individual, must be transferred to assure the generation-to-generation transmission of the specific content of reproductive invariance.

It will be readily seen that, in this or that species situated higher or lower on the animal scale, the achievement of the fundamental teleonomic project (i.e., invariant reproduction) calls assorted, more or less elaborate and complex structures and performances into play. The fact must be stressed that concerned here are not only the activities directly bound up with reproduction itself, but all those that contribute – be it very indirectly – to the species' survival and multiplication. For example, in higher mammals the play of the young is an important element of psychic development and social integration. Therefore this activity has teleonomic value, inasmuch as it furthers the cohesion of the group, a condition for its survival and for the expansion of the species. It is the degree of complexity of all these performances or structures, conceived as having the function of serving the teleonomic purpose, that we would like to estimate.

This magnitude, while theoretically definable, is not measurable in practice. Still, it may serve as a rule of thumb for ranking different species or groups upon a "teleonomic scale." To take an extreme example, imagine a bashful poet who, prevented by shyness from declaring his passion to the woman he loves, can only express it symbolically, in the poems he dedicates to her. Suppose that at last, conquered by these refined compliments, the lady surrenders to the poet's desire. His verses will have contributed to the success of his essential project, and the information they contain must therefore be tallied in the sum of the teleonomic performances assuring transmission of genetic invariance.

Indisputably, no analogous performance figures in the successful accomplishment of the project in other animal species, the mouse for instance. But—and this is the important point—the genetic invariance content is about the same in the mouse and the human being (and in all mammals, for that matter). *The two magnitudes we have been trying to define are therefore quite distinct.*

Which leads us to consider a most important question concerning the relationship among the three properties we singled out as characteristic of living beings. The fact that the computer program identified them successively and independently does not prove that they are not simply three manifestations of a single, more basic, more secret property, inaccessible to any direct observation. Were this so, the drawing of distinctions among the properties, the seeking of different definitions for them, might be nothing but delusion and arbitrariness. Far from shedding light on the real problem, far from tracking down "the secret of life" and truly dissecting it, we would be engaged merely in exorcizing it.

It is perfectly true that these three properties—teleonomy, autonomous morphogenesis, and reproductive invariance—are closely interconnected in all living beings. Genetic invariance expresses and reveals itself only through, and thanks to, the autonomous morphogenesis of the structure that constitutes the teleonomic apparatus.

There is this to be observed right away: not all of these three concepts have the same standing. Whereas invariance and teleonomy are indeed characteristic "properties" of living beings, spontaneous structuration ought rather to be considered a mechanism. Further on we shall see that this mechanism intervenes both in the elaboration of teleonomic structures and in the reproduction of invariant information as well. That it finally accounts for the latter two properties does not, however, imply that they should be regarded as one. It remains possible – it is in fact methodologically indispensable – to maintain a distinction between them, and this for several reasons:

1. One can at least *imagine* objects capable of invariant reproduction but devoid of any teleonomic apparatus. Crystalline structures offer one example of this, at a level of complexity admittedly very much lower than that of all known living organisms.

2. The distinction between teleonomy and invariance is more than a mere logical abstraction. It is warranted on grounds of chemistry. Of the two basic classes of biological macromolecules, one, that of proteins, is responsible for almost all teleonomic structures and performances; while genetic invariance is linked exclusively to the other class, that of nucleic acids.

3. Finally, as will be seen in the next chapter, this distinction is assumed, explicitly or otherwise, in all the theories, all the ideological constructions (religious, scientific, or philosophical) pertaining to the biosphere and to its relationship to the rest of the universe.

Living creatures are strange objects. At all times in the past, men must have been more or less confusedly aware of this. The development of the natural sciences beginning in the seventeenth century, their flowering in the nineteenth, instead of effacing this impression rather rendered it more acute. Over against the physical laws governing macroscopic systems, the very existence of living organisms seemed to constitute a paradox, violating certain of the fundamental principles modern science rests upon. Just which ones? That is not immediately clear. Hence the question is, precisely, to analyze the nature of this—or these—"paradoxes." This will give us occasion to specify the relative position, vis-à-vis physical laws, of the two essential properties that characterize living organisms: reproductive invariance and structural teleonomy.

Indeed at first glance invariance appears to constitute a profoundly paradoxical property, since the maintaining, the reproducing, the multiplying of highly ordered structures seems in conflict with the second law of thermodynamics. This law enjoins that no macroscopic system evolve otherwise than in a downward direction, toward degradation of the order that characterizes it.

the "paradox" of invariance

However, this prediction of the second law is valid, and verifiable, only if we are considering the overall evolution of an *energetically isolated* system. Within such a system, in one of its phases, we may see ordered structures take shape and grow without that system's overall evolution ceasing to comply with the second law. The best example of this is afforded by the crystallization of a saturated solution. The thermodynamics of such a system are well understood. The local enhancement of order represented by the assembling of initially unordered molecules into a perfectly defined crystalline network is "paid for" by a transfer of thermal energy from the crystalline phase to the solution: the entropy—or disorder—of the system as a whole augments to the extent stipulated by the second law.

This example shows that, within an isolated system, a local heightening of order is compatible with the second law. We have pointed out, however, that the degree of order represented by even the simplest organism is incomparably higher than that which a crystal defines. We must now ask whether the conservation and invariant multiplication of such structures is also compatible with the second law. This can be verified through an experiment closely comparable with that of crystallization.

We take a milliliter of water having in it a few milligrams of a simple sugar, such as glucose, as well as some mineral salts containing the essential elements that enter into the chemical constituents of living organisms (nitrogen, phosphorus, sulfur, etc.). In this medium we grow a bacterium, for example *Escherichia coli* (length, 2 microns; weight, approximately 5×10^{-13} grams). Inside thirty-six hours the solution will contain several billion bacteria. We shall find that about 40 per cent of the sugar has been converted into cellular constituents, while the remainder has been oxidized into carbon dioxide and water. By carrying out the entire experiment in a calorimeter, one can draw up the thermodynamic balance sheet for the operation and determine that, as in the case of crystallization, the entropy of the system as a whole (bacteria plus medium) has increased a little more than the minimum prescribed by the second law. Thus, while the extremely complex system represented by the bacterial cell has not only been conserved but has multiplied several billion times, the thermodynamic debt corresponding to the operation has been duly settled.

No definable or measurable violation of the second law

has occurred. Nonetheless, something unfailingly upsets our physical intuition as we watch this phenomenon, whose strangeness is even more appreciable than before the experiment. Why? Because we see very clearly that this process is bent or oriented in one exclusive direction: the multiplication of cells. These to be sure do not violate the laws of thermodynamics, quite the contrary. They not only obey them; they utilize them as a good engineer would, with maximum efficiency, to carry out the project and bring about the "dream" (as François Jacob has put it) of every cell: to become two cells.

Later we shall try to give an idea of the complexity, the subtlety, and the efficiency of the chemical machinery necessary to the accomplishment of a project demanding the synthesis of several hundred different organic constituents; their assembly into several thousand macromolecular species; and the mobilization and utilization, where necessary, of the chemical potential liberated by the oxidation of sugar: i.e., in the construction of cellular organelles. There is, however, no physical paradox in the invariant reproduction of these structures: invariance is bought at not one penny above its thermodynamic price, thanks to the perfection of the teleonomic apparatus which, grudging of calories, in its infinitely complex task attains a level of efficiency rarely approached by man-made machines. This apparatus is entirely logical, wonderfully rational, and perfectly adapted to its purpose: to preserve and reproduce the structural norm. And it achieves this, not by departing from physical laws, but by exploiting them to the exclusive advantage of its per-

teleonomy and the principle of objectivity

20

sonal idiosyncrasy. It is the very existence of this purpose, at once both pursued and fulfilled by the teleonomic apparatus, that constitutes the "miracle." Miracle? No, the real difficulty is not in the physics of the phenomenon; it lies elsewhere, and deeper, involving our own understanding, our intuition of it. There is, really, no paradox or miracle; but a flagrant *epistemological contradiction*.

The cornerstone of the scientific method is the postulate that nature is objective. In other words, the *systematic* denial that "true" knowledge can be got at by interpreting phenomena in terms of final causes—that is to say, of "purpose." An exact date may be given for the discovery of this canon. The formulation by Galileo and Descartes of the principle of inertia laid the groundwork not only for mechanics but for the epistemology of modern science, by abolishing Aristotelian physics and cosmology. To be sure, neither reason, nor logic, nor observation, nor even the idea of their systematic confrontation had been ignored by Descartes' predecessors. But science as we understand it today could not have been developed upon those foundations alone. It required the unbending stricture implicit in the postulate of objectivity—ironclad, pure, forever undemonstrable. For it is obviously impossible to imagine an experiment which could prove the *nonexistence* anywhere in nature of a purpose, of a pursued end.

But the postulate of objectivity is consubstantial with science; it has guided the whole of its prodigious development for three centuries. There is no way to be rid of it, even tentatively or in a limited area, without departing from the domain of science itself.

Objectivity nevertheless obliges us to recognize the teleonomic character of living organisms, to admit that in their structure and performance they act projectively – realize and pursue a purpose. Here therefore, at least in appearance, lies a profound epistemological contradiction. In fact the central problem of biology lies with this very contradiction, which, if it is only apparent, must be resolved; or else proven to be utterly insoluble, if that should turn out indeed to be the case.

II
Vitalisms and Animisms

SINCE THE TELEONOMIC properties of living beings appear to challenge one of the basic postulates of the modern theory of knowledge, any philosophical, religious, or scientific view of the world must, *ipso facto*, offer an implicit if not an explicit solution to this problem. Every solution

the priority relationship between invariance and teleonomy: a fundamental dilemma

in its turn, whatever the motivation behind it, just as inevitably implies a hypothesis as to the causal and temporal precedence, in relation to each other, of the two properties characteristic of living beings: invariance and teleonomy.

We shall defer until Chapter VI an exposition of, and justifications for, the single hypothesis that modern science here deems acceptable: namely, that invariance necessarily precedes teleonomy. Or, to be more explicit, the Darwinian idea that the initial appearance, evolution, and steady refinement of ever more intensely teleonomic structures are due to perturbations occurring in a structure *which already possesses the property of invariance* — hence is capable of preserving the effects of chance and thereby submitting them to the play of natural selection.

Of course the theory I am only briefly and dogmatically sketching here is not that of Darwin himself, who could not in his day have had any inkling of the chemical mechanisms of reproductive invariance, nor of the nature of the perturbations these mechanisms undergo. But it is no disparagement of Darwin's genius to note that the selective theory of evolution did not take on its full significance, precision, and certainty until less than twenty years ago.

Ranking teleonomy as a secondary property deriving from invariance – alone seen as primary – the selective theory is the only one so far proposed that is consistent with the postulate of objectivity. It is at the same time the only one not merely compatible with modern physics but based squarely upon it, without restrictions or additions. In short, the selective theory of evolution assures the epistemological coherence of biology and gives it its place among the sciences of "objective nature." A powerful argument indeed in favor of the theory; but less than enough to justify it.

All other concepts which have sought to provide an explicit answer to the problem of the strangeness of living beings, or which are implicitly contained in religious ideologies and most of the great philosophical systems, assume the reverse hypothesis: to wit, that *invariance is safeguarded, ontogeny guided*, and *evolution oriented* by an initial teleonomic principle, of which all these phenomena are the purported manifestations. I shall devote the remainder of this chapter to a schematic analysis of the logic of these interpretations, very diverse in appearance but all implying the renunciation, partial or total, admitted or not, conscious or otherwise, of the postulate of objectivity. It will be convenient here to classify these concepts (rather arbitrarily, it is true) under one of two headings, according to the nature and supposed extension of the teleonomic principle they invoke.

Thus on one side we may place a first group of theories involving a teleonomic principle which operates only within the biosphere, in the heart of "living matter." These theories, which I shall call *vitalist*, therefore imply a radical distinction between living beings and the inanimate world.

And on the other side we may group together the concepts that posit a *universal* teleonomic principle, responsible for the course of affairs throughout the cosmos as well as within the biosphere, where it is said to express itself simply in a more precise and intense manner. These theories see in living beings the most highly elaborated, most perfect products of a universally oriented evolution, which has culminated, because it had to, in man and mankind. These concepts I shall call *animist:* they are in many respects more interesting than the vitalist theories, to which I shall give only a cursory glance.*

Among vitalist theories a wide variety of tendencies may be discerned. Here we shall be content with distinguishing between what I shall refer to as "metaphysical vitalism" and "scientistic vitalism."

---

* It may be well to stress that I am here employing the qualifying "animist" and "vitalist" in a special sense, somewhat different from current usage.

Chance and Necessity

There has probably been no more illustrious proponent of a metaphysical vitalism than Henri Bergson. Thanks to

| metaphysical
| vitalism

an engaging style and a metaphorical dialectic bare of logic but not of poetry, his philosophy achieved immense success. It seems to have fallen into almost complete discredit today; but in my youth no one stood a chance of passing his baccalaureate examination unless he had read *Creative Evolution*. This philosophy, as some will recall, rests entirely upon a certain idea of life conceived as an *élan*, a "current," absolutely distinct from inanimate matter but contending with it, "traversing" it so as to force it into organized form. Contrary to almost all other vitalisms and animisms, that of Bergson predicates no ultimate goal: it refuses to put life's essential spontaneity in bondage to any kind of predetermination. Evolution, identified with the *élan vital* itself, can therefore have neither final nor efficient causes. Man is the supreme stage at which evolution has arrived, without having sought or foreseen it. He is rather the sign and proof of the total freedom of the creative *élan*.

This conception is bound up with another, considered by Bergson fundamental: rational intelligence is an instrument of knowledge specially designed for mastering inert matter but utterly incapable of apprehending life's phenomena. Only instinct, consubstantial with the *élan vital*, can give a direct, global insight into them. Every analytical and rational statement about life is therefore meaningless, or rather irrelevant. The high development of rational intelligence in Homo sapiens has brought on a grave and regrettable impoverishment of his powers of intuition, a lost treasure we today must strive to recover.

I shall not try to discuss this philosophy (which indeed does not lend itself to discussion). A captive of logic, and poor in global intuitions, I feel myself disqualified. Be that as it may, I do not regard Bergson's attitude as insignificant; quite the contrary. Conscious or unconscious rebellion against the rational, respect given to the *id* at the expense of the *ego*, are hallmarks of our times, and so too is creative spontaneity. Had Bergson employed a less limpid language, a more "profound" style, he would be reread today.*

The "scientific" vitalists have been more numerous, and they include some very distinguished scholars. But while fifty years ago the vitalists were recruited from among biologists (of whom the most renowned, Driesch, gave up embryology for philosophy), those of our day come mainly from the physical sciences, like Professors Elsässer and Polanyi. It is understandable, certainly, that physicists should be still more impressed than biologists by the strangeness of living things. Summarized in a few words, for example, here is Elsässer's position.

scientistic vitalism

The strange properties, invariance and teleonomy, are

---

* Bergson's thought, it need hardly be said, is not lacking in obscurity or patent contradictions. One may well question, for example, whether Bergsonian dualism is essential: should it not perhaps be seen as deriving from a more basic monism? (C. Blanchard, in a personal communication.) My intention here is, of course, not to explore Bergson's thought in its ramifications, but only in those implications which most directly concern the theory of living systems.

doubtless not at fundamental odds with physics; but the physical forces and chemical interactions brought to light by the study of nonliving systems *do not fully account for them.* Hence it must be realized that over and above physical principles *and adding themselves thereto*, others are operative in living matter, but not in nonliving systems where, consequently, these electively vital principles could not be discovered. It is these principles – or, to borrow from Elsässer's terminology, these "biotonic laws" – that must be elucidated.

Such hypotheses, it seems, were not dismissed by the great Nils Bohr himself. But he did not claim to have proof that they were necessary. Are they? That, finally, is the nub of it. That is what Elsässer and Polanyi assert. The least one can say is that the arguments of these physicists are oddly lacking in strictness and solidity.

These arguments concern respectively each of the strange properties. As regards invariance, its mechanism is sufficiently well known today for us to be able to state that no nonphysical principle is required for its interpretation.*

This leaves us with teleonomy or, more exactly, with the morphogenetic mechanisms which put teleonomic structures together. It is perfectly true that embryonic development is in appearance one of the most miraculous phenomena in the whole of biology. It is also true that these phenomena, admirably described by embryologists, continue in large part (for technical reasons) to elude genetic and biochemical analysis, obviously the sole avenue to an understanding of them. The attitude of the vital-

---

* See Chapter VI.

ists who feel that physical laws are – or in any case will prove themselves – insufficient to explain embryogenesis draws its justification, therefore, not from precise knowledge or from definite observations, but from our present-day ignorance alone.

On the other hand, our understanding of the molecular control mechanisms that regulate cellular growth and activity has progressed considerably and ought soon to contribute to the interpretation of organic development. We shall come to a discussion of these mechanisms in Chapter IV, and in that connection we shall have more to say about certain vitalist arguments. In order to survive as a point of view, vitalism requires that in biology there should remain, if not actual paradoxes, at least certain "mysteries." Developments in molecular biology over the past two decades have singularly narrowed the domain of the mysterious, leaving wide open to vitalist speculation little other than the field of subjectivity: that of consciousness itself. One runs no great risk in predicting that in this area as well, for the time being still "off limits," such speculation will prove just as sterile as in all the others where it has hitherto been practiced.

Animist conceptions, as I have already said, are in many respects a great deal more interesting than vitalist ideas. Reaching back to mankind's infancy, perhaps to before the appearance of Homo sapiens, they are still deep-rooted in the soul of modern man.

Our ancestors, we must presume, perceived the strangeness of their condition only very dimly. They did not have the reasons we have today for feeling themselves strangers in the universe upon which they opened their eyes. What did they see first? Animals, plants; beings

the animist projection and the "old covenant"

whose nature they could at once divine as similar to their own. Plants grow, seek sunlight, die; animals stalk their prey, attack their enemies, feed and protect their young; males fight for the possession of a female. About plants and animals as about man himself there was nothing hard to explain. These beings all have an aim, a purpose: to live and to go on living in their progeny, even at the price of death. Its purpose explains the being, and the being makes sense only through the purpose animating it.

But around them our ancestors also saw other objects, far more mysterious: rocks, rivers, mountains, the thunderstorm, the rain, the stars in the sky. If these objects exist it must also be for a purpose; to nourish it they had also to have a spirit or soul. Thus was the world's strangeness resolved for those early human beings: in reality there exist no inanimate objects. For such a thing would be incomprehensible. In the river's depths, on the mountaintop, more subtle spirits pursue vaster and more impenetrable designs than the transparent ones animating men and beasts. Thus were our forebears wont to see in nature's forms and events the action of forces either benign or hostile, but never indifferent—never totally alien.

Animist belief, as I am visualizing it here, consists essentially in a projection into inanimate nature of man's awareness of the intensely teleonomic functioning of his own central nervous system. It is, in other words, the hypothesis that natural phenomena can and must be explained in the same manner, by the same "laws," as subjective human activity, conscious and purposive. Primitive animism formulated this hypothesis with complete candor, frankness, and precision, populating nature with gracious or awesome myths and myth-figures which have for centuries nourished art and poetry.

One would be wrong to smile, even out of the fondness and deference the childlike inspire. Do we imagine that modern culture has really forsaken the subjective interpretation of nature? Animism established a covenant between nature and man, a profound alliance outside of which seems to stretch only terrifying solitude. Must we break this tie because the postulate of objectivity requires it? Ever since the seventeenth century the history of ideas attests to the profuse efforts put forth by the greatest minds to avert that break, to forge the old bond anew. Think of such mighty efforts as those of Leibnitz, or of the colossal and ponderous monument Hegel raised. But idealism has not by any means been the only refuge for a cosmic animism. At the very core of certain ideologies said and claiming to be founded upon science, the animist projection, in a more or less disguised form, turns up again.

The biological philosophy of Teilhard de Chardin would not merit attention but for the startling success it

scientistic progressism

has encountered even in scientific circles. A success which tells of the eagerness, of the need to revive the covenant. Teilhard revives it, and does so nakedly. His philosophy, like Bergson's, is based entirely upon an initial evolutionist postulate. But, unlike Bergson, he has the evolutive force operating throughout the entire universe, from elementary particles to galaxies: there is no "inert" matter, and therefore no essential distinction between "matter" and "life." His wish to present this concept as "scientific" leads Teilhard to base it upon a new definition of energy. This is somehow distributed between two vectors, one of which would be (I presume) "ordinary" energy, whereas the other would correspond to the upward evolutionary surge. The biosphere and man are the latest products of this ascent along the spiritual vector of energy. This evolution is to continue until all energy has become concentrated along the spiritual vector: that will be the attaining of "point omega."

Although Teilhard's logic is hazy and his style laborious, some of those who do not entirely accept his ideology yet allow it a certain poetic grandeur. For my part I am most of all struck by the intellectual spinelessness of this philosophy. In it I see more than anything else a systematic truckling, a willingness to conciliate at any price, to come to any compromise. Perhaps, after all, Teilhard was not for nothing a member of that order which, three centuries earlier, Pascal assailed for its theological laxness.

The idea of reestablishing the old animist covenant with nature, or of founding a new one through a universal theory according to which the evolution of the biosphere culminating in man would be part of the smooth onward flow of cosmic evolution itself—this idea did not of course originate with Teilhard. It is in fact the central theme of nineteenth-century scientistic progressism. One finds it at the very heart of Spencer's positivism and of the dialectical materialism of Marx and Engels as well. The unknown and *unknowable* force which, according to Spencer, operates throughout the universe creating variety, coherence, specialization, and order, plays what amounts to exactly the same role in Teilhard's "ascending" energy: human history is the extension of biological evolution, itself a component part of cosmic evolution. Thanks to this single principle, man at last finds his eminent and necessary place in the universe, along with certainty of the progress which is forever pledged to him.

Spencer's differentiating force, like Teilhard's ascending energy, is a plain instance of animist projection. In order to give meaning to nature, so that man need not be separated from it by a fathomless gulf, and for it again to become decipherable and intelligible, *a purpose had to be restored to it*. Should no spirit be available to harbor this purpose, then one inserts into nature an evolutive, an ascending "force," which in effect amounts to abandoning the postulate of objectivity.

Among the scientistic ideologies of the nineteenth century the most powerful, the one which in our times still wields a profound influence reaching far beyond the already vast circle of its adepts, is of course Marxism. Hence it is most interesting—and revealing—to note that, bent on grounding the edifice of their social doctrines upon a bedrock of nature's own laws, Marx and Engels resorted—they too, but more clearly and deliberately than Spencer—to "animist projection."

the animist projection in dialectical materialism

Indeed I do not see how else one can interpret the famous "inversion" by which Marx substitutes dialectical materialism for the idealist dialectic of Hegel.

Hegel's postulate, that the most general laws governing the universe in its evolution are of a dialectical order, is in its proper place within a system which acknowledges no permanent and authentic reality except mind. If all events, all phenomena are but partial manifestations of "an idea that thinks itself," it is legitimate to look for the most immediate expression of the universal laws in our subjective experience of the thinking process. And since thought proceeds dialectically, "the laws of the dialectic" govern the whole of nature. But to retain these subjective laws just as they are and make them serve as those of a purely material universe, this is to effect the animist projection in the most blatant manner and with all its consequences, the scrapping of the postulate of objectivity being the first.

Neither Marx nor Engels ventured a detailed analysis that would serve to justify the logic of this inversion of the dialectic. However, on the strength of the numerous examples of its application, given notably by Engels in his *Anti-Dühring* and *The Dialectics of Nature*, one may attempt to reconstruct the underlying thought of the founders of dialectical materialism. Its essential tenets would be these:

1. Movement is the mode of existence of matter.

2. The universe, defined as the totality of matter, which is all that exists, is in a state of perpetual evolution.

3. All true knowledge of the universe is of the kind that contributes to the intelligence of this evolution.

4. But this knowledge is obtained only in the interaction, itself evolutive and a cause of evolution, between man and matter (or, more exactly, the "rest" of matter). All true knowledge is therefore "practical."

5. Consciousness pertains to this cognitive interaction. Conscious thought consequently reflects the movement of the universe itself.

6. Since, then, thought is a part and reflection of universal movement, and since its movement is dialectical, the evolutionary law of the universe itself must be dialectical. Which explains and justifies the use of such terms as *contradiction, affirmation*, and *negation* in connection with natural phenomena.

7. The dialectic is constructive (thanks notably to the "third law"). Therefore the evolution of the universe is itself ascendant and constructive. Its highest expression is human society, consciousness, thought, all necessary products of this evolution.

8. Through its stressing of the evolutionary essence of the universe's structures, dialectical materialism goes far beyond the materialism of the eighteenth century which, founded upon classical logic, was limited to recognizing only mechanical interactions between supposedly invariant objects, and therefore remained incapable of evolutionary thinking.

To be sure, one may contest this reconstruction and deny that it reflects Marx and Engels' authentic thought. But that is really only secondary. The influence of an ideology depends upon the meaning it maintains in the minds of its adepts, and which is spread by later commentators. Countless texts show that the foregoing summary is legitimate, as representing at least the "vulgate" of dialectical materialism. One example will do, especially significant inasmuch as its author, J. B. S. Haldane, was an outstanding modern biologist. He writes in his preface to the English translation of *The Dialectics of Nature:*

Marxism has a twofold bearing on science. In the first place Marxists study science among other human activities. They show how the scientific activities of any society depend on its changing needs, and so in the long run on its productive methods, and how science changes the productive methods, and therefore the whole of it. But secondly Marx and Engels were not content to analyze the changes in society. In dialectics they saw the science of the general laws of change, not only in society and in human thought, but in the external world which is mirrored by human thought. That is to say it can be applied to problems of "pure" science as well as to the social relations of science.*

The external world "mirrored by human thought": there indeed is the gist of it. The logic of the inversion obviously requires that this mirroring be a good bit more than a by and large faithful transposition of the external world. For dialectical materialism it is indispensable that the *Ding an sich*—the thing or phenomenon in itself—reach the level of consciousness unaltered and undiminished, with none of its properties suppressed. The external world in the whole fullness and integrity of its structures and movement must be literally present to consciousness.†

---

* Friedrich Engels, *The Dialectics of Nature.*

† From Henri Lefebvre *(Le Matérialisme dialectique* [Paris: PUF, 1949], p. 92) we take the following passage: "Far more than a mere process of thought, dialectic exists prior to mind, inheres in being. It obtrudes itself upon mind. First we analyze the simplest stirring of thought; of the most abstract, the barest thought. In so doing we discover the most general categories and their concatenation. These we must next connect to the concrete movement, to the *given content;*

No doubt certain of Marx' own writings could be cited in opposition to this concept. For all that, it remains indispensable to the logical coherence of dialectical materialism, as later Marxists, if not Marx and Engels themselves, were to realize very well. Let us not forget, moreover, that dialectical materialism is a relatively late adjunct to the socioeconomic edifice Marx had already raised. An adjunct that was clearly intended to make of historical materialism a "science" based upon the laws of nature itself.

Their insistence upon the "perfect mirror" explains the dialectical materialists' dogged repudiation of any kind of critical epistemology, immediately condemned out of hand as "idealist" or "Kantian." To be sure, this attitude is to some extent understandable on the part of men living in the nineteenth century, contemporary witnesses to the first great scientific upheaval. It might then very well look as though, thanks to science, man was in the way of achieving direct mastery over nature, appropriating its very substance. Nobody doubted, for instance, that gravitation was one of the laws of nature itself, probed to its furthermost depths.

the need for a critical epistemology

As we know, it was by a return to the sources—the sources of knowledge itself—that the spadework was done for the second age of science, that of the twentieth

---

we are then made aware of the fact that the process that involves the content and the self clarifies itself for us in the workings of the laws of dialectic. Contradictions in thought come not from thinking alone, from its weaknesses or incoherence; they come also from the content. Their interlocking tends toward the expression of the *total movement of the content* and lifts it to the level of consciousness and of reflection."

century. By the close of the nineteenth it once again becomes evident that a critical epistemology is absolutely necessary; indeed, that the objectivity of knowledge is contingent upon it. From now on this critique is the concern not only of philosophers but of scientists as well, who are led to incorporate it into the texture of theory itself. Only thus could the theory of relativity and quantum mechanics be developed.

Moreover, advances in neurophysiology and in experimental psychology are presently beginning to disclose at least some aspects of the functioning of the nervous system. Enough to make it clear that the information the central nervous system furnishes to consciousness is only, and probably can only be, in codified form, transposed, framed within preexisting norms: in other words, assimilated and not just restored.

The pure reflection thesis, of the perfect mirror which would not even invert the image, must therefore strike us as still less defensible than it did our grandparents. But after all, a perceptive eye hardly needed what twentieth-century science was to bring, in order to see the confusions and foolishness to which this notion was bound to lead. To set straight poor Herr Dühring, an early recalcitrant, Engels himself proposed numerous examples of the dialectical interpretation of natural phenomena. There is the memorable one of the grains of barley, to illustrate the "Third Law."

the epistemological bankruptcy of dialectical materialism

Let us take a grain of barley. . . . If such a grain of barley meets with conditions which are normal for it, if it falls on suitable soil, then under the influence

of heat and moisture it undergoes a specific change,
it germinates; the grain as such ceases to exist, it is
negated, and in its place appears the plant which has
arisen from it, the negation of the grain. But what is
the normal life-process of this plant? It grows, flow-
ers, is fertilized and finally once more produces
grains of barley, and as soon as these have ripened
the stalk dies, is in its turn negated. As a result of this
negation of the negation we have once again the original grain of barley, but not as a single unit, but ten-,
twenty- or thirty-fold. . . .

It is the same [Engels adds a little further on] in mathematics. Let us take any algebraic quantity whatever:
for example, a. If this is negated, we get $-a$ (minus a).
If we negate that negation, by multiplying $-a$ by $-a$,
we get $+a^2$, i.e., the original positive quantity, but at a
higher degree, raised to its second power. . . .

And so forth.

What these examples illustrate above all is the scope of
the epistemological disaster that ensues from the "scientific" use of dialectical interpretations. Modern dialectical
materialists ordinarily manage to avoid such silliness. But
to make dialectical contradiction the "fundamental law"
of all movement, all evolution, is nonetheless to try to
systematize a subjective interpretation of nature whereby
it may be shown to have an ascending, constructive, creative intent, a purpose; in short, to render nature decipherable and morally meaningful. This is "animist projection"
again, always recognizable whatever its disguises.

This interpretation is not only foreign to science but
incompatible with it—and as such it has appeared every
time the dialectical materialists, emerging from purely

"theoretical" verbiage, have sought to use their ideas to help light the path of experimental science. Although he had a thorough acquaintance with the science of his day, Engels himself had been led to reject, in the name of dialectics, two of the greatest discoveries of the age: the second law of thermodynamics and (notwithstanding his admiration for Darwin) the theory of natural selection. It was by virtue of these same principles that Lenin assailed the epistemology of Mach; that, later, Zhdanov ordered Russian thinkers to scourge the Copenhagen school for "its devilish Kantian mischief"; that Lysenko accused geneticists of maintaining a theory radically at odds with dialectical materialism, and therefore necessarily false. Despite the disclaimers of the Russian geneticists, Lysenko was perfectly right: the theory of the gene as the hereditary determinant, invariant from generation to generation and even through hybridizations, is indeed completely irreconcilable with dialectical principles. It is by definition an idealist theory, since it rests upon a postulate of invariance. The fact that today the structure of the gene and the mechanism of its invariant reproduction are known does not redeem anything, for modern biology's description of them is purely mechanistic. And so, at best, they are concepts ascribable to "vulgar materialism," mechanistic, and hence "objectively idealist," as M. Althusser pointed out in his severe commentary upon my inaugural lecture before the Collège de France.

I have reviewed these various ideologies or theories briefly and only very incompletely. Some may find that I have given a distorted, because partial, image of them. For this I shall try to excuse myself by reminding the reader that I have not sought here to go further than to sift out

Vitalisms and Animisms

the anthropocentric illusion

what these concepts hold, or imply, with respect to biology, and more especially the relationship they assume between invariance and teleonomy. It has been seen that, without exception, all take an initial teleonomic principle as the *primum movens* of evolution, whether of the biosphere alone or of the entire universe. In the eyes of modern scientific theory all these concepts are erroneous, not only for reasons of method (since in one way or another they imply abandonment of the postulate of objectivity) but for factual reasons, which will be discussed below.

At the source of these errors lies, of course, the anthropocentric illusion. The heliocentric theory, the concept of inertia, and the principle of objectivity were never enough to dissipate that ancient mirage. Rather than dispelling the illusion, the theory of evolution at first seemed to endow it with a new reality by making of man no longer the center of the entire universe but its natural heir, awaited from time immemorial. God could at last die, replaced by this new and grandiose fantasy. The ultimate aim of science from now on would be to formulate a unified theory which, based on a small number of principles, would account for the whole of reality, biosphere and man included. It was this exalting certainty that constituted the fare upon which nineteenth-century scientistic progressism fed. A unified theory which, for their part, the dialectical materialists believed they had already formulated.

Because he saw it as jeopardizing the certainty that man and human thought are necessary end-products of a cosmic progress, Engels felt constrained to deny the second law. It is significant that he does so right in the introduction to *The Dialectics of Nature,* and that he moves directly from this subject to an impassioned cosmological prediction by which he promises eternal recurrence, if not to the human species, at any rate to the "thinking mind." A recurrence indeed, but of one of mankind's most ancient myths.*

Not until the second half of this century was the new anthropocentric illusion, propped up on the theory of evolution, to give way in its turn. I believe that we can assert today that a universal theory, however completely successful in other domains, could never encompass the biosphere, its structure, and its evolution as phenomena *deducible* from first principles.

**the biosphere
a unique occurrence
nondeducible
from first
principles**

---

* "Hence," Engels declares, "we arrive at the conclusion that in some way, which it will later be the task of scientific research to demonstrate, the heat radiated into space must be able to become transformed into another form of motion, in which it can once more be stored up and rendered active. Thereby the chief difficulty in the way of the reconversion of extinct suns into incandescent vapour disappears. . . .

"But however often, and however relentlessly, this cycle is completed in time and space, however many millions of suns and earths may arise and pass away, however long it may last before the conditions for organic life develop, however innumerable the organic beings that have to arise and to pass away before animals with a brain capable of thought are developed from their midst, and for a short span of time find conditions suitable for life, only to be exterminated later without mercy, we have the certainty that matter remains eternally the same in all its transformations, that none of its attributes can ever be lost, and therefore, also, that with the same iron necessity that it will exterminate on earth its highest creation, the thinking mind, it must somewhere else and at another time again produce it" (*The Dialectics of Nature*).

This proposition may appear obscure. Let us try to make it clearer. A universal theory would obviously have to extend to include relativity, the theory of quanta, and a theory of elementary particles. Provided certain initial conditions could be formulated, it would also contain a cosmology which would forecast the general evolution of the universe. We know however (contrary to what Laplace believed, and after him the science and "materialist" philosophy of the nineteenth century) that these predictions could be no more than statistical. The theory might very well contain the periodic table of elements, but could only determine the probability of existence of each of them. Likewise it would anticipate the appearance of such objects as galaxies or planetary systems, but in no case could it deduce from its principles the necessary existence of this or that object, event, or individual phenomenon—whether it be the Andromeda nebula, the planet Venus, Mount Everest, or yesterday evening's thundershower.

In a general manner the theory would anticipate the existence, the properties, the interrelations of certain *classes* of objects or events, but would obviously not be able to foresee the existence or the distinctive characteristics of any *particular* object or event.

The thesis I shall present in this book is that the biosphere does not contain a predictable class of objects or of events but constitutes a particular occurrence, compatible indeed with first principles, but not *deducible* from those principles and therefore essentially unpredictable.

Let there be no misunderstanding here. In saying that as a class living beings are not predictable upon the basis of first principles, I by no means intend to suggest that they are not *explicable* through these principles—that they transcend them in some way, and that other principles, applicable to living systems alone, must be invoked. In my view the biosphere is unpredictable for the very same reason—neither more nor less—that the particular configuration of atoms constituting this pebble I have in my hand is unpredictable. No one will find fault with a universal theory for not affirming and foreseeing the existence of this particular configuration of atoms; it is enough for us that this actual object, unique and real, be *compatible* with the theory. This object, according to the theory, is under no obligation to exist; but it has the right to.

That is enough for us as concerns the pebble, but not as concerns ourselves. We would like to think ourselves necessary, inevitable, ordained from all eternity. All religions, nearly all philosophies, and even a part of science testify to the unwearying, heroic effort of mankind desperately denying its own contingency.

III
Maxwell's Demons

THE CONCEPT of teleonomy implies the idea of an *oriented, coherent,* and *constructive* activity. By these standards proteins must be deemed the essential molecular agents of teleonomic performance in living beings.

proteins as molecular agents of structural and functional teleonomy

1. Living beings are chemical machines. The growth and multiplication of all organisms require the accomplishing of thousands of chemical reactions whereby the essential constituents of cells are elaborated. This is what is called "metabolism." It is organized along a great number of divergent, convergent, or cyclical "pathways," each comprising a sequence of reactions. The precise adjustment and high efficiency of this enormous, yet microscopic chemical activity are maintained by a certain class of proteins, the enzymes, playing the role of specific catalysts.

2. Like a machine, every organism, down to the very "simplest," constitutes a coherent and integrated functional unit. Plainly enough, the functional coherence of so complex a chemical machine, which is autonomous as well, calls for a cybernetic system governing and controlling the chemical activity at numerous points. Especially

as regards the higher organisms, we are still a long way from elucidating the entire structure of these systems. Nevertheless a great many of its elements are known at present, and in all these cases it turns out that its essential agents are so-called "regulatory" proteins which act, in effect, as detectors of chemical signals.

3. The organism is a self-constructing machine. Its macroscopic structure is not imposed upon it by outside forces. It shapes itself autonomously by dint of constructive internal interactions. Although our understanding of the mechanisms of development is still more than imperfect, we are now in a position to state that the constructive interactions are microscopic and molecular, and that the molecules involved are essentially if not uniquely proteins.

Hence they are proteins which channel the activity of the chemical machine, assure its coherent functioning, and put it together. All these teleonomic performances rest, in the final analysis, upon the proteins' so-called "stereospecific" properties, that is to say upon their ability to "recognize" other molecules (including other proteins) by their *shape*, this shape being determined by their molecular structure. At work here is, quite literally, a microscopic discriminative (if not "cognitive") faculty. We may say that any teleonomic performance or structure in a living being—whatever it may be—can, in principle at least, be analyzed in terms of stereospecific interactions involving one, several, or a very large number of proteins.*

---

* I have deliberately oversimplified here. Certain DNA structures play a role that must be considered teleonomic. And certain RNAs

It is on the structure, on the shape of a given protein that the particular stereospecific discrimination constituting its function depends. To the extent that we could retrace the origin and evolution of this structure we would also be describing the origin and evolution of the teleonomic performance it discharges.

In the present chapter we shall discuss the specific catalytic function of proteins; in the following one, their regulatory function; and in Chapter V their constructive function. The problem of the origin of functional structures will be taken up in this chapter and further dealt with in the next.

One may indeed study the functional properties of a protein without having to refer to the details of its particular structure. (Actually, to date we have a thoroughly detailed knowledge of the three-dimensional structure of only some fifteen proteins.) A few general facts need recalling, however.

Proteins are very large molecules, in molecular weight ranging from 10,000 to 1,000,000 or more. These macromolecules are constituted by the sequential polymerization of components whose molecular weight is about 100, and which belong to the class of amino acids. Every protein thus contains from 100 to 10,000 amino acid residues. These very numerous residues belong, however, to

(ribonucleic acids) constitute essential elements of the machinery which translates the genetic code (cf. Appendix 3, p. 193). However, particular proteins are also involved in these mechanisms which, at nearly every stage, bring interactions between proteins and nucleic acids into play. Discussion of these mechanisms may be omitted without affecting the analysis of teleonomic molecular interactions and their general interpretation.

only twenty different chemical species,* which are encountered in all living beings, from bacteria to man. This sameness of composition is one of the most striking illustrations of the fact that the prodigious diversity of *macroscopic* structures of living beings rests in fact on a profound and no less remarkable unity of *microscopic* make-up. About this we shall have more to say.

On the basis of their overall shape, one may divide proteins into two main classes:

a. The so-called "fibrous" proteins are very elongated molecules which in living beings play a principally mechanical role, like the rigging on a sailing vessel; although the properties of some of these proteins (those found in muscle) are of great interest, we must refrain from discussing them here.

b. The so-called "globular" proteins are by far the more numerous and, through their functions, the more interesting; in these proteins the strands constituted by the sequential polymerization of amino acids fold upon themselves in an exceedingly complex manner, thereby giving these molecules a compact, pseudo-globular shape.

Even the simplest organisms contain a very great number of different proteins. It may be put at 2500 ± 500 for the bacterium *Escherichia coli* (weighing 5×10^{-13} grams and 2 microns in length, approximately). For the higher mammals such as man, one may suggest a figure on the order of a million.

Among the thousands of chemical reactions that contribute to the development and performances of an organ-

---

* See Appendix 1, p. 184.

the enzyme-proteins as specific catalysts

ism, each is provoked electively by a particular enzyme-protein. Oversimplifying only a very little, one may say that in the organism each enzyme exerts its catalytic activity at but one single point in the metabolism. It is above all through the extraordinary *electivity* of action they display that enzymes differ from nonbiological catalysts used in the laboratory or in industry. Some of the latter are exceedingly active – capable, in very slight quantity, of greatly accelerating various reactions. However, not one of these catalysts comes near the meanest ordinary enzyme for specificity of action.

This specificity is twofold:

1. Each enzyme catalyzes but one type of reaction.

2. Among the sometimes very numerous compounds in the organism susceptible of undergoing that type of reaction, the enzyme, as a general rule, is active in regard to only one.

A few examples will help clarify these propositions. There exists an enzyme, called fumarase, which catalyzes the hydration of fumaric acid into malic acid. This reaction, diagrammed below, is reversible; that is, the same enzyme also catalyzes the dehydration of malic acid into fumaric acid.

$$
\begin{array}{ccc}
\text{COOH} & & \text{COOH} \\
| & & | \\
\text{CH} & \xrightleftharpoons{\;(+\,H_2O)\;} & \text{HCOH} \\
\| & & | \\
\text{CH} & & \text{CH}_2 \\
| & & | \\
\text{COOH} & & \text{COOH} \\
\textbf{(fumaric acid)} & & \textbf{(malic acid)}
\end{array}
$$

49

Meanwhile there exists a geometric isomer of fumaric acid, maleic acid –

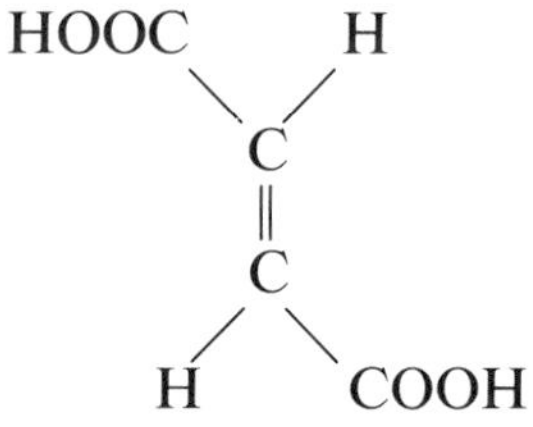

$$HOOC \diagdown C \diagup H$$

(fumaric acid)

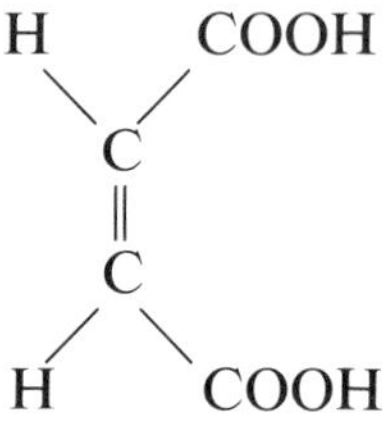

(maleic acid)

chemically capable of undergoing the same hydration. The enzyme is totally inactive with regard to the second.

But there exist as well two *optical* isomers of malic acid, which possesses an asymmetric carbon:*

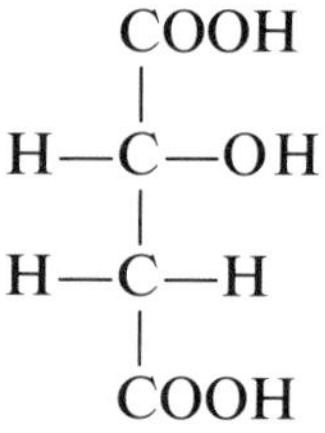

$$
\begin{array}{c}
COOH \\
| \\
H-C-OH \\
| \\
H-C-H \\
| \\
COOH
\end{array}
$$

(L-malic acid)

$$
\begin{array}{c}
COOH \\
| \\
HO-C-H \\
| \\
H-C-H \\
| \\
COOH
\end{array}
$$

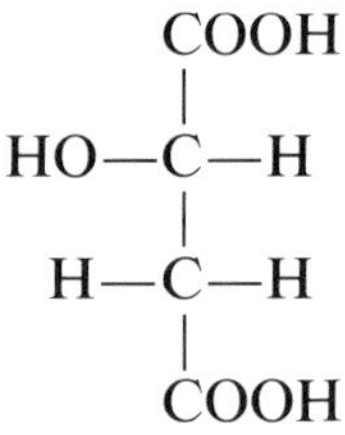

(D-malic acid)

Mirror-images of each other, these two compounds are chemically equivalent and practically inseparable by classical chemical techniques. Between the two the enzyme nevertheless exercises an absolute discrimination.

---

* Compounds consisting of a carbon atom linked to four different groupings are thereby deprived of symmetry. They are said to be "optically active," for the passage of polarized light through such compounds imparts to the plane of polarization a rotation to the left (L: levogyrous compounds) or to the right (D: dextrogyrous compounds).

Thus,

a. The enzyme dehydrates L-malic acid exclusively in order to produce fumaric acid exclusively; and

b. Starting with fumaric acid, the enzyme produces L-malic acid exclusively, but not D-malic acid.

The rigorous distinction the enzyme makes between optical isomers is more than just a striking illustration of the *steric* specificity of enzymes. Here, to begin with, one finds the explanation for the previously mysterious fact that among the many chemical cellular constituents that are asymmetric (the case with the majority of them) only one of the two optical isomers is, as a general rule, represented in the biosphere.

But, in the second place, according to Curie's very general principle governing the conservation of symmetry, the fact that from an optically symmetrical compound (fumaric acid) an asymmetrical compound is obtained enjoins that

a. It be the enzyme itself that constitutes the "source" of the asymmetry; hence, that it be itself optically active, which indeed it is; and

b. The substrate's initial symmetry is lost in the course of its interaction with the enzyme-protein. The hydration reaction, then, must occur within a "complex" formed by a temporary association between enzyme and substrate; in such a complex the initial symmetry of fumaric acid would be effectively lost.

"Stereospecific complex" as accounting for the specificity as well as for the catalytic activity of enzymes – the concept is of key importance. We shall return to it after discussing some further examples.

There exists in certain bacteria another enzyme, called aspartase, which also acts upon fumaric acid alone, to the

51

exclusion of every other compound and notably of its geometric isomer, maleic acid. The reaction of "addition upon a double bond" catalyzed by this enzyme is a close analogy to the preceding one. This time it is not a molecule of water but of ammonia that is condensed with fumaric acid to give an amino acid, aspartic acid:

(fumaric acid)   (L-aspartic acid)

Aspartic acid possesses an asymmetrical carbon atom; it is therefore optically active. As in the preceding case, the enzymatic reaction produces exclusively one of the isomers, the one of the L series, called a "natural" isomer because amino acids entering into the composition of proteins all belong to the L series.

The two enzymes aspartase and fumarase thus discriminate strictly, not only between the optical and geometrical isomers of their substrates and products, but likewise between the molecules of water and ammonia. One is led to say that these latter molecules also enter into the composition of the stereospecific complex, within whose framework the addition reaction is produced; and that in this complex the molecules are rigorously positioned each in respect to the other. Both the specificity of action and the stereospecificity of the reaction would seem to result from this positioning.

From the foregoing examples the existence of a stereospecific complex, as intermediary in enzymatic reactions, could be deduced only as an explanatory hypothesis. But in certain favorable cases the existence of this complex may be demonstrated directly. One such case is that of the enzyme called β-galactosidase, which specifically catalyzes the hydrolysis of compounds possessing the structure labeled (A) in the diagram below.

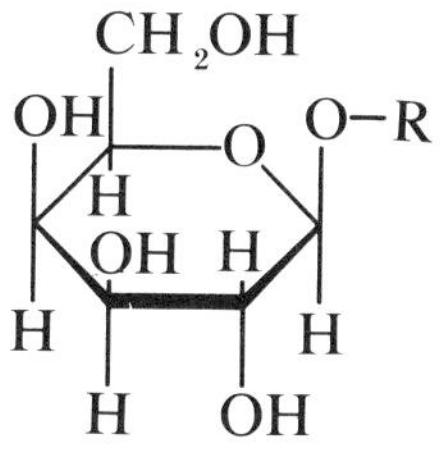

(A) 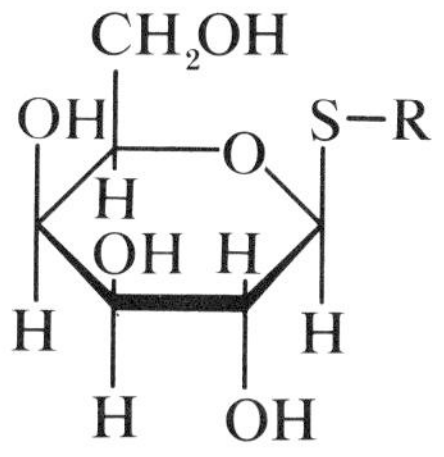　　　　　　　　(B)

Let us bear in mind that there exist many isomers of such compounds (sixteen geometrical isomers, differing by the relative orientation of the OH and H groupings upon carbons 1 to 5; plus the optical opposites of each of these sixteen).

The enzyme in point of fact exactly discriminates between all these isomers, and hydrolyzes only one of them. Nevertheless one may "trick" the enzyme by synthesizing "steric analogues" of compounds belonging to this series, in which the oxygen of the hydrolyzable bond is replaced by sulfur—formula (B) in the above diagram. The sulfur atom, larger than the oxygen, is of the same valency, and for both atoms the orientation of the valencies is the same. The three-dimensional *shape* of these sulfur derivatives is therefore practically the same as that of their oxygen counterparts. But the bond formed by sulfur is much more stable than the oxygen bond. The enzyme consequently fails to hydrolyze these compounds. That they however form a stereospecific complex with the protein may be *directly* demonstrated.

Such observations not only confirm the theory of the complex but show that an enzymatic reaction is to be considered as made up of two distinct steps:

1. The formation of a stereospecific complex between protein and substrate.

2. The catalytic activation of a reaction within the complex: a reaction *oriented* and *specified* by the structure of the complex itself.

This distinction is of major importance, and will enable us to arrive at one of the central concepts of molecular biology. But first we must note that among the different

covalent and noncovalent bonds

types of bonds which contribute to the stability of a chemical edifice, two classes are to be made out: the covalent bond and the noncovalent. Covalent bonds—often referred to under the name of "chemical bonds" *sensu stricto*—are due to the sharing of electronic orbitals between two or several atoms; noncovalent bonds to several other types of interaction not implying the sharing of electronic orbitals.

It is not necessary, for our main purposes here, to

dwell upon the nature of the physical forces that take part in these different types of interaction. We may start by emphasizing that the two classes of bonds differ from each other through the energy of the associations they ensure. Simplifying somewhat, and specifying that we are now considering only those reactions occurring in aqueous phase, we may say that the average amount of energy absorbed or liberated by a reaction involving covalent bonds is on the order of 5 to 20 Kcal per bond. For a reaction involving noncovalent bonds only, the average amount of energy would be between 1 and 2 Kcal.*

This considerable difference partially accounts for the difference in stability between "covalent" and "noncovalent" chemical constructs. The essential, however, lies not there but in the difference in the so-called "activation" energies brought into play in the two types of reaction. This point is of highest importance. To clarify it we should be reminded that a reaction causing a molecular population to pass from a given stable state into another must be understood to include an intermediate state, of potential energy higher than that of either of the two terminal states. The process is often represented by a plot whose abscissa indicates the forward course of the reaction and its ordinate the potential energy (Fig. 1). The difference in potential energy between the terminal states

---

* Let us remember that a bond's energy is, by definition, the energy that must be furnished *to split it*. But in actual fact most chemical — and notably biochemical — reactions consist in the *exchange* of bonds rather than in their outright rupture. The energy brought into play in a reaction is that which corresponds to an exchange of the type AY + BX→AX + BY. It is therefore always lower than the splitting energy.

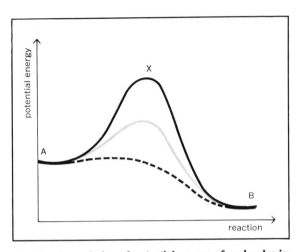

Fig. 1. Diagram showing the variation of potential energy of molecules in the course of a reaction. A: initial stable state; B: final stable state; X: intermediate state of potential energy superior to that of the two stable states. Continuous line: covalent reaction; gray line: covalent reaction in the presence of a catalyzer that lowers the activation energy; dashes: noncovalent reaction.

corresponds to the energy released by the reaction; the difference between the initial state and the intermediate ("activated") state is the activation energy: this is the energy that the molecules must *transitorily* acquire in order to enter into reaction. This energy, acquired in the course of the first phase and released in the second, does not figure in the final thermodynamic accounting. However, upon it depends the *speed* of the reaction, which will be practically zero at ordinary temperature, if the activation energy is high. Hence, in order to provoke such a reaction one must either considerably increase the temperature (thereby assuring sufficient energy to the molecules) or else employ a catalyst, whose role is to "stabilize" the activated state, thereby reducing the difference of potential between this state and the initial one.

Now – and this is the crucial point – in general:

a. The activation energy of covalent reactions is high; their speed is therefore very slow or zero at low temperature and in the absence of catalysts; while

b. The activation energy of noncovalent reactions is very low if not zero; they therefore occur *spontaneously* and *very rapidly*, at *low temperature*, and in the *absence of catalysts*.

The result is that structures defined by noncovalent interactions can attain a certain stability only if they entail *multiple* interactions. Furthermore, noncovalent interactions acquire a notable amount of energy only when the atoms lie a very short distance apart, practically "touching" one another. Consequently two molecules (or areas of molecules) will be able to contract a noncovalent association only if the surfaces of both include *complementary sites* permitting several atoms of the one to enter into contact with several atoms of the other.

If we now add that the complexes formed between enzyme and substrate are of a noncovalent nature it will be

the concept of the noncovalent stereospecific complex

seen why these complexes are *necessarily* stereospecific: they can form only if the enzyme molecule has a site exactly "complementary" to the shape of the substrate molecule. It will be seen, also, that in the complex the molecule of substrate is necessarily very strictly positioned by virtue of the multiple interactions connecting it to the enzyme molecule's receptor site.

And, lastly, it will be seen that depending upon the *number* of noncovalent interactions it entails, the stability of a noncovalent complex will vary along a very broad scale. Therein lies a precious property of noncovalent complexes: their stability can be exactly adjusted to the function fulfilled. Enzyme-substrate complexes must be able to assemble and to come apart very rapidly; high catalytic activity demands this. These complexes are indeed easily and very promptly dissociable. Other complexes, whose function is permanent, acquire a stability of the same order as that of a covalent association.

Until now we have discussed only the first step in an enzymatic reaction: the forming of the stereospecific complex. The catalytic step itself, which follows formation of the complex, need not long detain us for, from the biological viewpoint, it poses no such deeply significant problems as the preceding one. The belief today is that enzymatic catalysis results from the inductive and polarizing action of certain chemical groupings present in the protein's "specific receptor." Aside from specificity (due to the substrate molecule's very precise positioning vis-à-vis the inducer groups), the catalytic effect is explained by schemes similar to those which account for the action of nonbiological catalysts (such as, notably, H^+ and OH^- ions).

The formation of the stereospecific complex, as a prelude to the catalytic act itself, may therefore be regarded as simultaneously fulfilling two functions:

1. The exclusive *choice* of a substrate, determined by its steric structure.

2. The correct *presentation* of the substrate in the precise position that limits and specifies the catalytic effect of the inducer groups.

Maxwell's Demons

The idea of a noncovalent stereospecific complex is applicable not just to enzymes nor even, as will be seen, just to proteins. It is of pivotal importance for the interpretation of all the phenomena of choice, of elective discrimination, that characterize living beings and make them appear to escape the fate spelled out by the second law of thermodynamics. In this connection it is worth glancing again at the example of fumarase.

Using organic chemistry's means to aminate fumaric acid, one obtains a mixture of the two optical isomers of aspartic acid. The enzyme, on the other hand, catalyzes exclusively the formation of L-aspartic acid. This represents an input of information exactly corresponding to a binary choice (since there are two isomers). Here one sees at the most elementary level how structural information can be created and distributed in living beings. The enzyme of course possesses, in the structure of its specific receptor, the information corresponding to this choice. But the energy required to *amplify* this information does not come from the enzyme: to orient the reaction exclusively along one of the two possible paths, the enzyme utilizes the chemical potential constituted by the fumaric acid solution. All the synthesizing activity of cells, however complex, may in the last analysis be interpreted in the same terms.

These phenomena, prodigious in their complexity and their efficiency in carrying out a preset program, clearly invite the hypothesis that they are guided by the exercise of somehow "cognitive" functions.

Maxwell's demon

Chance and Necessity

The nineteenth-century physicist James Maxwell attributed such a function to his microscopic demon. We recall how this hypothetical personage, posted at the communicating opening between two enclosed spaces filled with a gas of whatever kind, was supposed, without any consumption of energy, to maneuver an ideal hatch enabling him to prevent certain molecules from passing from one chamber to the other. The gatekeeper could thus "choose" to allow only fast (high energy) molecules through in one direction, and only slow (low energy) molecules in the other. The result being that, of the two enclosed spaces originally at the same temperature, one grew hotter while the other grew cooler – all without any apparent consumption of energy. However imaginary this experiment, it caused physicists no end of perplexity: for it did indeed seem that *through the exercise of his cognitive function* the demon was able to violate the second law. And as this cognitive function appeared neither measurable nor even definable from the physical standpoint, Maxwell's "paradox" seemed to defy all analysis in operational terms.

The key to the riddle was provided by Léon Brillouin, drawing upon earlier work by Szilard: he demonstrated that the exercise of his cognitive function by the demon had *necessarily* to entail the consumption of a certain amount of energy which, on balance, precisely offset the lessening entropy within the system as a whole. So as to work the hatch "intelligently," the demon must first have *measured* the speed of each particle of gas. Now any reckoning – that is to say, any acquisition of information – presupposes an interaction, in itself energy-consuming.

This famous theorem is one of the sources of modern

thinking regarding the equivalence between information and negative entropy. The theorem interests us here precisely because enzymes, at the microscopic level, exercise an order-creating function. But this creation of order, as we have seen, is not gratuitous; it comes about at the expense of a consumption of chemical potential. In short, the enzymes function exactly in the manner of Maxwell's demon corrected by Szilard and Brillouin, draining chemical potential into the processes chosen by the program of which they are the executors.

Let us retain the essential idea developed in this chapter: it is by virtue of their capacity to form, with other molecules, *stereospecific and noncovalent complexes* that proteins exercise their "demoniacal" functions. The following chapters will illustrate the crucial importance of this key concept, which will recur as the ultimate interpretation of the most distinctive properties of living beings.

IV
Microscopic Cybernetics

BY VIRTUE OF its extreme specificity, an "ordinary" enzyme (like those taken as examples in the previous chapter) constitutes a completely independent functional unit. The "cognitive" function of the "demons" is restricted to the recognition of their specific substrate, to the exclusion both of all other compounds and of anything that may occur within the cell's chemical machinery.

From a glance at a drawing condensing what is now known of cellular metabolism we can tell that even if at

functional coherence of cellular machinery

each step each enzyme carried out its job perfectly, the sum of their activities could only be chaos were they not somehow interlocked so as to form a coherent system. We do indeed have the most manifest evidence of the extreme efficiency of the chemical machinery of living beings, from the "simplest" to the most complex.

We have of course long been aware of the existence in animals of systems providing large-scale coordination of the organism's performances: that is what the nervous and endocrine systems do. These systems insure coordination between organs and tissues; which is to say, finally, *among cells*. And we now know that within each cell a cybernetic network hardly less (if not still more) complex guarantees the functional coherence of the intracellular chemical machinery—this is what has emerged from studies dating back only twenty years, some to but five or ten.

We are still far short of having analyzed in its entirety the system that governs the metabolism, growth, and division of bacteria, the simplest

regulatory proteins and the logic of regulations

known cells. But thanks to thorough analysis of certain parts of this system, we understand fairly well today the principles it works by. It is these principles we shall be discussing in this chapter. We shall see that the elementary control operations are handled by specialized proteins acting as detectors and transducers of chemical information.

At the present time the best known of these regulatory proteins are the so-called "allosteric" enzymes. They compose a special class, by reason of features which distinguish them from "ordinary" enzymes. Like the latter, allosteric enzymes recognize and bind electively a particular substrate and activate its conversion into products. But these enzymes have the further property of recognizing electively one or several *other* compounds, whose (stereospecific) association with the protein has a modifying effect—that is, depending upon the case, of *heightening or inhibiting its activity with respect to the substrate.*

The regulatory, coordinating function of interactions of this type (known as allosteric interactions) stands proven today by countless examples. These interactions may be classified into a certain number of "regulatory patterns," depending upon the relationship existing between the reaction in question and the metabolic origin of the "allosteric effectors" controlling it. The main regulatory patterns are these (Fig. 2):

1. *Feedback inhibition.* The enzyme which catalyzes the first reaction of a sequence whose end-product is an essential metabolite (a constituent of proteins or of nucleic acids, for example*) is inhibited by the final product of the sequence. The intracellular concentration of this metabolite therefore governs its own rate of synthesis.

2. *Feedback activation.* The enzyme is activated by a product of degradation of the terminal metabolite. This case is frequent with metabolites whose high chemical potential constitutes a source of energy for the cellular machinery. This regulatory pattern hence contributes to maintaining the available chemical potential at a prescribed level.

3. *Parallel activation.* The first enzyme of a metabolic sequence leading to an essential metabolite is activated by a metabolite synthesized by an independent and parallel sequence. This mode of regulation contributes to maintaining a balance between metabolites belonging to the same family and destined for assembly in one of the classes of macromolecules.

4. *Activation through a precursor.* The enzyme is activated by a compound which is a more or less remote precursor of its immediate substrate. This mode of regulation amounts to keeping the "demand" subordinate to the "of-

---

* Any compound produced by metabolism is called a metabolite; "essential metabolites" are the compounds universally required for the growth and multiplication of cells.

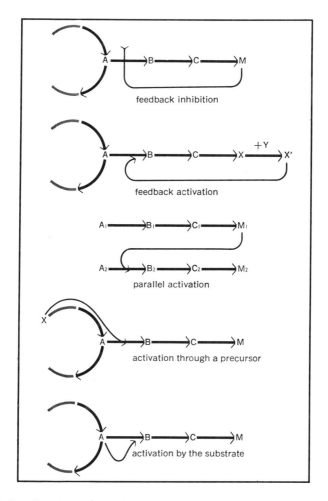

feedback inhibition

feedback activation

parallel activation

activation through a precursor

activation by the substrate

Fig. 2. Various "regulatory modes" assured by allosteric interactions. Arrows with solid lines symbolize reactions producing intermediate compounds (denoted A, B, etc.). The letter M represents the terminal metabolite, conclusion of the sequence of reactions. Fine lines indicate the origin and point of application of a metabolite acting as an allosteric effector, the inhibitor or activator of a reaction.

65

fer." One particular but extremely frequent case of this kind is

5. *Activation* of the enzyme *by the substrate* itself. This then plays its own "ordinary" role and at the same time that of an allosteric effector with respect to the enzyme.

Rarely is an allosteric enzyme subject to only one mode of regulation. As a general rule these enzymes are under the simultaneous control of several allosteric effectors, antagonistic or cooperative. A frequently encountered situation is a "ternary" regulation comprising:

a. Activation by the substrate (pattern 5);

b. Inhibition by the end-product of a sequence (pattern 1); and

c. Parallel activation by a metabolite of the same family as the end-product (pattern 3).

Here, then, the enzyme simultaneously recognizes all three effectors and "measures" their relative concentrations; its activity at any time represents a summing up of these three inputs of information.

To illustrate the refined intricacy of these systems we may mention by way of example the regulatory patterns of "branching" metabolic pathways (Fig. 3), which are numerous. In these cases, in general, not only are the initial reactions, at the metabolic fork, regulated by feedback inhibition, but an earlier reaction, higher up on the common branch, is cogoverned by the two (or several) final metabolites.* The danger of blocking the synthesis

---

* E. R. Stadtman, *Advances in Enzymology, 28* (1966), 41–159. G. N. Cohen, *Current Topics in Cellular Regulation, 1* (1969), 183–231.

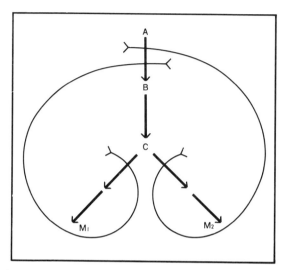

Fig. 3. Allosteric regulation of branching metabolic pathways. For the meaning of symbols (letters and arrows) see Fig. 2.

of one of the metabolites by an excess of the other is skirted, depending upon the particular case, in one of two different ways, either

a. By delegating to this one reaction two distinct enzymes, each inhibited by one of the metabolites to the exclusion of the other; or

b. With a single enzyme, which is only inhibited by the two metabolites acting "in concert" but not by either one of them alone.

The fact must be underlined that, leaving aside the substrate, the effectors which regulate an allosteric enzyme's activity take no part in the reaction itself. With the enzyme they usually form a noncovalent complex, entirely and instantaneously reversible, from which they come away completely unmodified. The consumption of energy incidental to the regulatory interaction is practically nil: it represents but a tiny fraction of the effectors' intracellular chemical potential. On the other hand the catalytic reaction governed by these very weak interactions may, for its part, involve relatively considerable energy transfers. These systems, thus, are comparable to those employed in electronic automation circuitry, where the very slight energy consumed by a relay can trigger a large-scale operation, such as, for example, the firing of a ballistic missile.

Just as an electronic relay can be controlled simultaneously by several electric potentials, so, as we have seen, an allosteric enzyme is ordinarily controlled by several chemical potentials. But the analogy goes still further. As is well known, it is usually advantageous to have a relay system respond *nonlinearly* to the variations in the potential governing it; threshold effects are thus obtained, permitting finer regulation. The same holds true in the case of most allosteric enzymes. The curve showing the variation of activity as dependent upon concentration of an effector (including the substrate) is almost always *S*-shaped. In other words the effect of the ligand* at first increases *faster* than its concentration. This behavior is the more remarkable in that it appears to be characteristic of allosteric enzymes. In ordinary, "classic" enzymes, on the contrary, the effect always increases *more slowly* than the concentration.

I am not sure what the minimal weight might be for an

---

* We give the name of "ligand" to a compound defined by its ability to bind to another specific compound.

electronic relay presenting the same logical features as an average allosteric enzyme (receiving and integrating inputs from three or four sources, and responding with threshold effect). Let us say something like a hundredth of a gram. The weight of an allosteric enzyme molecule capable of the same performances is of the order of 10^{-17} of a gram. Which is a million billion times less than an electronic relay. That astronomical figure affords some idea of the "cybernetic" (i.e., teleonomic) power at the disposal of a cell equipped with hundreds or thousands of these microscopic entities, all far more clever than the Maxwell-Szilard-Brillouin demon.

The question is to understand how an allosteric protein performs these extraordinary feats. It is known now that allosteric interactions are mediated by discrete shifts in the protein's molecular shape. In the next chapter we shall see that a globular protein's involuted and compact form is stabilized by a host of noncovalent bonds which together cooperate in maintaining its structure. This allows certain proteins to assume two (or more) conformational states (just as certain compounds may exist in different allotropic states). The two states in question, and the "allosteric transition" wherein the molecule shifts from one to the other or back again, are often symbolized thus:

mechanism of allosteric interactions

$$(R) \quad\quad\quad\quad (T)$$

Since the shape-recognizing properties of a protein depend upon the shape of its binding site or sites, it may be asserted (and in certain favorable instances directly demonstrated) that these stereospecific properties are modified by the transition. For example: in state R the protein will be able to recognize and therefore to bind compound α at one site (but not compound β), whereas in state T it will recognize and bind compound β but not α. It follows that either compound will have the effect of stabilizing the protein in this or that of its two states, R or T, at the expense of the other; and that α and β will be mutually antagonistic, since their respective interactions with the protein are mutually exclusive. Imagine now a third compound γ (which could be the substrate) binding exclusively with the R state in some other site of the molecule than the one where α binds: α and γ, it will be seen, cooperate in stabilizing the protein in its active state (the state which recognizes the substrate). Compound α and substrate γ will therefore function as activators, compound β as an inhibitor. The activity of a population of molecules will be proportional to the fraction of them that are in state R, a fraction which will be larger or smaller depending upon the relative concentration of the three ligands as well as upon the value of the intrinsic equilibrium between R and T. Thus the catalytic reaction will be controlled by the magnitude of these three chemical potentials.

Here let us emphasize what is by far the most important implication of the foregoing: namely, that the cooperative or antagonistic interactions of the three ligands are *totally indirect. There are no actual interactions between the ligands themselves; all the interactions occur exclusively between the protein and each ligand separately.* Further on we shall return to this idea, the apparently indispensable key to an understanding of the origin and development of cybernetic systems in living beings.*

This scheme of indirect interactions enables us to account as well for the subtle refinement evidenced in the protein's "nonlinear" response to variations in the concentration of its effectors. All known allosteric proteins are "oligomeric," made up by the noncovalent assembly of a few (often two or four; less often six, eight, or twelve) chemically identical subunits or "protomers." Each protomer bears a receptor for each of the ligands the protein recognizes. As a consequence of its assembly with one or several other protomers the steric structure of each of them is partially "constrained" by its neighbors. But theory, confirmed by crystallographic evidence, tells us that oligomeric proteins tend to "pack" in such a way—to adopt such structures—that all the protomers are geometrically equivalent; the constraints they are subjected to are therefore symmetrically distributed between protomers.

Let us now take the simplest case, that of a dimer, and consider what ensues from its dissociation into two monomers. Asunder, they are free to assume a "relaxed" state, structurally different from the constrained one into which each had been forced when linked together.

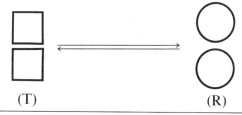

(T)   (R)

---

* J. Monod, J.-P. Changeux, and F. Jacob, *Journal of Molecular Biology,* 6 (1963), 306–29.

The two protomers' change of state, we shall say, is "concerted." It is this acting in concert that explains the nonlinearity of response: a ligand molecule's stabilization of the dissociated state R in one of the monomers prevents the other from returning to the associated state, and the same applies for a shift in the opposite direction. The equilibrium between the two states will be a quadratic function of the concentration of the ligands: it would be a fourth-power function for a tetramer, and so on.*

I have deliberately confined myself to discussing the simplest possible model, the operations I have described being present in certain systems which we have reason to regard as "primitive." In real systems the dissociation is only rarely complete: the protomers remain associated in both states, though more loosely in one of them. Upon this basic theme there are many possible variations, but the essential point has been to demonstrate that molecular mechanisms, extremely simple in themselves, account for the "integrative" properties of allosteric proteins.

Each of the allosteric enzymes referred to up until now constitutes a unit fulfilling a chemical function and at the same time a mediating element in regulatory interactions. Their properties give us an insight into how the homeostatic state of cellular metabolism is maintained at a peak of efficiency and coherence.

But by "metabolism" we essentially mean the transformations of small molecules and the mobilization of chem-

---

* J. Monod, J. Wyman, and J.-P. Changeux, *Journal of Molecular Biology, 12* (1965), 88–118.

ical potential. In cellular chemistry syntheses proceed upon yet another level: that of the macromolecules, nucleic acids and proteins (including, notably, the enzymes themselves). It has been known for quite some time that regulatory systems function at this level also. Their study is much more difficult than that of allosteric enzymes, and as a matter of fact only one of them, thus far, has been thoroughly analyzed. It shall be taken as an example.

This system, called the "lactose system," governs the synthesis of three proteins in the bacterium *Escherichia coli*. One of these proteins, galactoside permease, enables the galactosides* to penetrate and accumulate within the cell—whose membrane, in the absence of this protein, is impermeable to these sugars. A second protein hydrolyzes the β-galactosides. As for the third protein, its function is not altogether clear and is probably minor. The first and second, on the other hand, are both simultaneously indispensable to the metabolic utilization of lactose (and other galactosides) by the bacteria.

When *Escherichia coli* bacteria grow in a medium devoid of galactosides the three proteins are synthesized at an exceedingly slow rate: about one molecule every five generations. Almost immediately (within about two minutes) after a galactoside—in this connection termed an "inducer"—is added to the medium, the rate of synthesis of all three proteins increases a thousandfold and maintains this pace as long as the inducer is present. Let the inducer be withdrawn, and inside two or three minutes

---

* See Chapter III, p. 53.

Chance and Necessity

the rate of synthesis slips back to what it was initially.

The conclusions of a long study of this wonderfully and almost miraculously teleonomic phenomenon are summarized in Fig. 4.* Here we need not discuss the

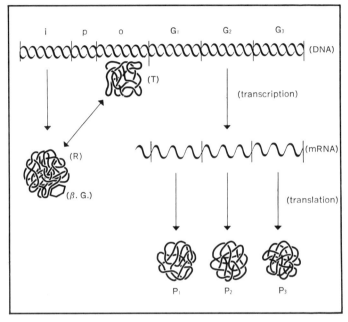

Fig. 4.   Regulation of the synthesis of the enzymes in the "lactose system."
R: repressor-protein, in state of association with the galactoside inducer shown by the hexagon.
T: repressor-protein in state of association with operator segment (o) of DNA.
i: "regulator gene" governing synthesis of the repressor.
p: "promoter" segment, point of initiation for synthesis of messenger RNA (m RNA).
G_1, G_2, G_3: "structure" genes governing synthesis of the three proteins in the system, marked P_1, P_2, P_3. (See text, p. 75.)

---

* The Finnish scientist Karstrom, who in the thirties made notable contributions to the study of these phenomena, later gave up research, apparently in order to become a monk.

right-hand part of the diagram, which represents the operation of messenger-RNA synthesis and its translation into polypeptide sequences. Let us simply observe that since the messenger has a rather brief existence (it lives for only a few minutes), its rate of synthesis determines the three proteins' rate of synthesis. Our chief interest is in the components of the regulatory system. They are:

the "regulator" gene, i

the "repressor" protein, R

the "operator" segment of the DNA, o

the DNA "promoter" segment, p

a molecule of inducer galactoside, βG

Its functioning is as follows:

1. The regulator gene directs the synthesis, at a constant and very slow rate, of the repressor protein.

2. The repressor specifically recognizes the operator segment to which it binds, with it forming a very stable complex (corresponding to a ΔF of some 15 Kcal).

3. In this state, synthesis of messenger (implying the intervention of the enzyme RNA polymerase) is blocked, presumably by simple steric hindrance, the beginning of this synthesis having to occur on the level of the promoter.

4. The repressor also recognizes β-galactosides, but binds them firmly only when in a free state: hence, in the presence of β-galactosides the operator-repressor complex is dissociated, this permitting the synthesis of messenger and consequently of protein.*

It is important to note that both interactions of the re-

---

* F. Jacob and J. Monod, *Journal of Molecular Biology, 3* (1961), 318–56. See also *The Lactose Operon*, Cold Spring Harbor Monograph, ed. J. R. Beckwith and David Zipser (1970).

pressor are noncovalent and reversible, and that, in particular, the inducer is not modified through its binding to the repressor. Thus the logic of this system is simple in the extreme: the repressor inactivates transcription; it is inactivated in its turn by the inducer. From this double negation results a positive effect, an "affirmation." The logic of this negation of the negation, we may add, is not dialectical: it does not result in a new statement but in the reiteration of the original one, written within the structure of DNA in accordance with the genetic code. The logic of biological regulatory systems abides not by Hegelian laws but, like the workings of computers, by the propositional algebra of George Boole.

A great many other similar systems (in bacteria) are known to us today. Not so far has any single one been taken down entirely. It appears very likely, though, that the logic of some of them is more complicated than that of the lactose system; in some, for example, negative interactions do not exclusively prevail. But the most general and most significant conclusions to be drawn from the analysis of the lactose system apply as well to these others. Of them all it may be said that

a. The repressor, having no activity of its own, is purely a transducer – a mediator – of chemical signals.

b. The effect of galactoside upon enzyme synthesis is totally indirect, due exclusively to the repressor's recognition properties and to the fact that two states, each exclusive of the other, are accessible to it. Here again we have what may be termed an allosteric interaction in the general sense discussed earlier.

c. There is no *chemically necessary* relationship between the fact that β-galactosidase hydrolyzes β-galactosides, and the fact that its biosynthesis is induced by the same compounds. Physiologically useful or "rational," this relationship is chemically arbitrary – "gratuitous," one may say.

This fundamental concept of *gratuity* – i.e., the independence, chemically speaking, between the function itself and the nature of the chemical signals controlling it – applies to allosteric enzymes. In this case one and the same protein molecule does double duty as specific catalyst and as transducer of chemical signals. But, as we have seen, allosteric interactions are indirect, proceeding exclusively from the protein's discriminatory properties of stereospecific recognition, in the two (or more) states accessible to it. Between the substrate of an allosteric enzyme and the ligands prompting or inhibiting its activity there exists no chemically necessary relationship of structure or of reactivity. The specificity of the interactions, in short, has nothing to do with the structure of the ligands; it is entirely due, instead, to that of the protein in the various states it is able to adopt, a structure in its turn freely, arbitrarily *dictated* by the structure of a gene.

the concept of gratuity

From this it results – and we come to our essential point – that so far as regulation through allosteric interaction is concerned, *everything is possible.* An allosteric protein should be seen as a specialized product of molecular "engineering," enabling an interaction, positive or negative, to come about between compounds without chemical affinity, and thereby eventually subordinating any reaction to the intervention of compounds that are chemically foreign and indifferent to this reaction. The

77

way in which allosteric interactions work hence permits a complete freedom in the "choice" of controls. And these controls, having no chemical requirements to answer to, will be the more responsive to physiological requirements, and will accordingly be selected for the extent to which they confer heightened coherence and efficiency upon the cell or organism. In a word, the very gratuitousness of these systems, giving molecular evolution a practically limitless field for exploration and experiment, enabled it to elaborate the huge network of cybernetic interconnections which makes each organism an autonomous functional unit, whose performances appear to transcend the laws of chemistry if not to ignore them altogether.*

Actually, as we have seen, when analyzed at the microscopic – the molecular – level, these performances appear wholly interpretable in terms of specific chemical interactions, electively assured, freely chosen and organized by regulatory proteins. And it is in the structure of these molecules that one must see the ultimate source of the autonomy, or more precisely, the self-determination that characterizes living beings in their behavior.

The systems we have studied up to this point are among those that coordinate the cell's activity and make a functional unit of it. In pluricellular organisms the coordination between cells, tissues, or organs is guaranteed by specialized systems: not only the nervous and endocrine systems, but also direct interactions between cells. I shall not discuss the functioning of these systems, of which we have as yet only the bare beginnings of a microscopic

---

* J. Monod, J.-P. Changeux, and F. Jacob, *Journal of Molecular Biology*, 6 (1963), 306 – 29.

description. It will be our hypothesis, however, that in these systems the molecular interactions which ensure the transmission and interpretation of chemical signals rest upon proteins endowed with discriminatory stereospecific recognition properties, and to which the essential principle of chemical gratuity applies, as we have seen it does in the case of allosteric interactions.

To end this chapter, a few words might perhaps be said about the old quarrel between "reductionists" and "holists." Certain schools of thought (all more or less consciously or confusedly influenced by Hegel) challenge the value of the *analytical* approach to systems as complex as living beings. According to these holist schools which, phoenixlike, spring up anew with every generation,* only failure awaits attempts to reduce the properties of a very complex organization to the "sum" of the properties of its parts. A most foolish and wrongheaded quarrel it is, merely testifying to the "holists'" profound misappreciation of scientific method and of the crucial role analysis plays in it. If a Martian engineer were trying to understand one of our earthling computers, how far could he conceivably get were he, on principle, to refuse to dissect the basic electronic components which in the machine execute the operations of propositional algebra? If any one branch of molecular biology illustrates better than others the sterility of holist theses as against the cogency of analytical meth-

"holism"
vs. **"reductionism"**

---

* Cf. Koestler and Smythies, ed., *Beyond Reductionism* (London: Hutchinson, 1969).

od, it is indeed the study of these microscopic cybernetic systems, at which we have taken a brief look in this chapter.

The analysis of allosteric interactions reveals, first of all, that teleonomic performances are not the exclusive endowment of complex, multicomponent systems, since *a* protein molecule shows itself capable, not only of electively activating a reaction, but of regulating its activity in response to input emanating from *several* chemical sources.

Secondly, thanks to the concept of gratuity, we see how and why these molecular regulatory interactions, foiling chemical constraints, succeed in being selectively chosen solely on the grounds of their contribution to the coherence of the system.

Lastly, from the study of these microscopic systems we come to see that for complexity, for richness, for potency, the cybernetic network in living beings far surpasses anything that the study of the overall behavior of whole organisms could ever hint at. And even though these analyses are not yet near to furnishing us with a complete description of the cybernetic system of the simplest cell, they tell us that, without exception, all the activities that contribute to the growth and multiplication of that cell are interconnected and intercontrolled, directly or otherwise.

On such a basis, but not on that of a vague "general theory of systems,"* it becomes possible for us to grasp in what very real sense the organism does effectively transcend physical laws—even while obeying them—thus achieving at once the pursuit and fulfillment of its own purpose.

---

* Von Bertalanffy, *ibid.*

V
Molecular
Ontogenesis

IN BOTH their macroscopic structure and their functions, living beings, as we have seen, are closely comparable to machines. On the other hand, they differ radically from them in their manner of coming into being. A machine — any artifact — owes its macroscopic structure to the action of external forces, of tools which impose shape upon matter. It is the sculptor's chisel that elicits the form of Aphrodite from the block of marble; as for the goddess herself, she was born of sea-foam (impregnated by the blood from Uranus' mutilated genitals), whence her body rose of itself, by itself.

In this chapter I wish to show that this process of spontaneous and autonomous morphogenesis rests, at bottom, upon the stereospecific recognition properties of proteins; that it is primarily a microscopic process before manifesting itself in macroscopic structures. Finally, it is the primary structure of proteins that we shall consult for the "secret" to those cognitive properties thanks to which, like Maxwell's demons, they animate and build living systems.

Let it be said at the outset that the problems we are about to tackle, those of the mechanisms of development, contain enigmas to which biology still has no answer. For while embryologists have provided admirable descriptions of development, we are a long way yet from knowing how to analyze the ontogenesis of macroscopic structures in terms of microscopic interactions. Nonetheless, the construction of certain molecular edifices is today fairly well understood, and the construction process, as I shall try to show, is veritably one of "molecular ontogenesis" in which the physical essence of the phenomenon becomes apparent.

As I indicated earlier, globular protein molecules often appear in the form of aggregates containing a definite number of chemically identical subunits. This number being usually small, these proteins have been designated as "oligomers." In oligomers the subunits (protomers) are associated by noncovalent bonds. Moreover, as we have already seen, the arrangement of the protomers within the oligomeric molecule is such that each of them is geometrically equivalent to the others. Each protomer, consequently, may be converted into any one of the others by an operation of symmetry — actually by a rotation. It is easily demonstrated that the oligomers so constituted possess the elements of symmetry of one of the rotational point-groups.

Thus these molecules constitute real microscopic crystals. They belong, however, to a special class which I shall call "closed crystals" for, contrary to ordinary crystals (whose geometry conforms to one of the so-called space-groups), they cannot grow without acquiring new elements of symmetry, while usually shedding some of those they have.

the spontaneous association of subunits in oligomeric proteins

In Chapter IV we spoke of how certain of the functional properties of these proteins are connected with their oligomeric state, including their symmetrical structure. The problem of how these microscopic edifices are constructed is hence quite as significant from the biological viewpoint as it is interesting from that of physics.

Since the protomers in an oligomeric molecule generally are associated by noncovalent bonds only, it is often possible to separate them into free monomeric units by relatively mild treatment (involving no recourse, for example, to high temperatures or aggressive chemical agents). In this state the protein will in general have lost all its functional properties, catalytic or regulatory. However – and this is the important point – if the initial "normal" conditions are restored (by eliminating the dissociating agent), the subunits will ordinarily reassemble spontaneously, re-forming the original "native" state of the aggregate: the same number of protomers in the same geometrical arrangement, accompanied by the same functional properties as before.

What is more, the reassembly of subunits belonging to a given species of protein will occur not only in a solution containing solely that particular protein, but also and just as well in complex "soups" made up of hundreds or thousands of other proteins. Which is further proof of the existence of an extremely specific recognition process, obviously due to the formation of noncovalent steric complexes interassociating the protomers. This process may be justly considered *epigenetic** since, out of a solution of monomeric molecules devoid of any symmetry,

---

* The appearance of new structures and new properties in the course of embryonic development has often been referred to as an "epigenetic" process, expressive of the gradual enrichment of the organism as it grows from its bare genetic beginnings represented by the initial

larger and more distinctly ordered molecules have appeared, and have come forth with functional properties hitherto completely absent.

Here what chiefly interests us is the *spontaneous* character of this molecular process of epigenesis. A twofold spontaneity:

1. The chemical potential necessary for the forming of the oligomers does not have to be injected into the system: it must be considered to be present in the solution of monomers.

2. Spontaneous in the thermodynamic sense, the process is also kinetically spontaneous: no catalyst is required to activate it—this, of course, because the bonds formed are noncovalent. We have already insisted upon the importance of the fact that both the formation and the splitting of such bonds involve next to nothing in the way of energy.*

Such a phenomenon is strictly comparable to molecular crystallization occurring in a solution of component molecules. There, too, order is constituted spontaneously through the interassociation of molecules belonging to a single chemical species. The analogy is further reinforced when, in either case,

the spontaneous structuration of complex particles

---

egg. The adjective is also often employed in reference to now outmoded theories in which the "preformationists" (who believed that the egg contained a miniature of the adult animal) were lined up in opposition to the "epigeneticists" (who believed in an *actual* enrichment of the initial genetic information. The term is employed here, not in relation to any theory, but in reference to all processes of structural and functional development.

* See Chapter III, p. 56.

structures arranged according to simple and repetitive geometric rules are seen to take shape. But it has been recently shown that certain organelles much more complex in structure are also the products of spontaneous assembly. This is the case with particles called ribosomes which are the essential components of the mechanism that translates the genetic code, that is, of the protein-synthesizing machinery. These particles, whose molecular weight attains 10^6, are made up by the assembly of some thirty distinct proteins plus three different types of nucleic acids. Although we do not know exactly how these various constituents are disposed within a ribosome, it is certain that their arrangement is extremely precise and that the functioning of the particle depends upon it. Now it has been found that, *in vitro*, the dissociated constituents of ribosomes spontaneously reassemble themselves into particles having the same composition, the same molecular weight, the same functional activity as the original "native" material.*

However, the most spectacular example we know of so far of the spontaneous construction of complex molecular edifices is without doubt that of certain bacteriophages.†
The complicated and very precise structure of the T4 bacteriophage corresponds to this particle's function, which is not only to protect the genome (i.e., the DNA) of the virus, but to attach itself to the wall of the host cell in order to inject into it, syringelike, its DNA content. The different parts of this microscopic precision machinery

---

* M. Nomura, "Ribosomes," *Scientific American, 221* (October, 1969), 28.
† Viruses which attack bacteria.

can be obtained separately from different mutants of the virus. Mixed together *in vitro* they assemble themselves *spontaneously* to reconstitute particles identical to normal ones and fully capable of exercising their DNA-injecting function.*

All these findings are relatively recent, and in this area of research we may anticipate important advances leading to the *in vitro* reconstitution of more and more complex organelles, such as mitochondria and membranes. The two or three cases just spoken of suffice, however, to illustrate the process whereby complex structures possessing functional properties develop from the stereospecific, spontaneous assembling of their protein constituents. Order, structural differentiation, acquisition of functions—all these appear out of a random mixture of molecules individually devoid of any activity, any intrinsic functional capacity other than that of recognizing the partners with which they will build the structure. And while in connection with ribosomes and bacteriophages we can no longer speak of crystallization, since these particles are of a degree of complexity, that is to say of an order, much higher than that of a crystal, in the last analysis it is nonetheless true that the chemical interactions involved are basically of the same nature as those that construct a molecular crystal. As in a crystal, the structure of the assembled molecules itself constitutes the source of "information" for the construction of the whole. These epigenetic processes therefore consist essentially

---

* R. S. Edgar and W. B. Wood, "Morphogenesis of bacteriophage T4 in extracts of mutant infected cells," *Proceedings of the National Academy of Science,* 55 (1966), 498.

in this: the overall scheme of a complex multimolecular edifice is contained *in posse* in the structure of its constituent parts, but only comes into actual existence through their assembly.

This analysis plainly reduces the old dispute between preformationists and epigeneticists to a quibbling over words. No preformed and complete structure preexisted anywhere; but the architectural plan for it was present in its very constituents. It can therefore come into being spontaneously and autonomously, without outside help and without the injection of additional information. The necessary information was present, but unexpressed, in the constituents. The epigenetic building of a structure is not a *creation*; it is a *revelation*.

Though admitting that the extrapolation still needs the support of conclusive experimental evidence, modern biologists are convinced that this concept, directly founded upon study of the formation of microscopic edifices, also explains and must be applied to the epigenesis of macroscopic structures (tissues, organs, limbs, etc.). Indeed, when we move on to macroscopic structures the problems, in terms both of dimensions and of complexity, are on a very different scale. Here the most important constructive interactions occur not between molecular components but between cells. That isolated cells of a given tissue are able to recognize one another discriminatively and to associate is an established fact; but the com-

microscopic morphogenesis and macroscopic morphogenesis

ponents or structures permitting cells to identify each other are still unknown. Everything suggests that the answer is to be sought in the structural characteristics of cellular membranes. Yet it remains uncertain whether the recognition is of individual molecular shapes or of multi-molecular surface patterns.* Whatever the case may be, and even if it is one of patterns not made up of protein components alone, the structure of patterns such as these would of necessity be determined by the shape-recognition properties of their protein components, and by those also of the enzymes responsible for the biosynthesis of a pattern's other components (polysaccharides or lipids, for example).

And so it may be that the "cognitive" properties of cells are not the direct but rather an exceedingly indirect expression of the discriminatory faculties of certain proteins. Nevertheless the construction of a tissue or the differentiation of an organ—macroscopic phenomena— must be viewed as integrated results of multiple microscopic interactions due to proteins, and as deriving from the stereospecific recognition properties belonging to those proteins, by way of the *spontaneous* forming of noncovalent complexes.

But we must hasten to say that this "reduction to the microscopic" of morphogenetic phenomena does not as yet constitute a working theory of those phenomena. Rather, it simply sets forth the principle in whose terms such a theory would have to be formulated if it were to

---

* J.-P. Changeux, in "Symmetry and function in biological systems at the molecular level," *Nobel Symposium* No. 11, ed. A. Engström and B. Strandberg, New York (1969), pp. 235–56.

aspire to anything better than simple phenomenological description. This principle defines the objective to be reached but furnishes little in the way of clues of how to get there. Only consider the magnitude of the problem of accounting in molecular terms for the elaboration of an apparatus as intricate as the central nervous system, requiring billions of specific interconnections between cells, some of them lying at relatively great distances apart in the body.

This problem of long-distance influences and orientations is probably the most difficult and the most important in embryology. In their efforts to explain the phenomena of regeneration, embryologists have introduced the idea of a "morphogenetic field" or "gradient," which at first glance seems a great leap beyond a stereospecific molecular interaction within the narrow confines of a few angstroms. However, the latter alone makes precise physical sense, and it is by no means inconceivable that a row of such interactions, one triggering the next, could create or define an organization of, for example, millimetric or centimetric proportions. In modern embryology the thinking is along these lines. It is fairly likely that the idea of purely *static* stereospecific interactions will turn out to be insufficient for the interpretation of the morphogenetic field or gradients. It will need the reinforcement of kinetic hypotheses, similar perhaps to those that make possible the interpretation of allosteric interactions. But I for my part remain convinced that only the shape-recognizing and stereospecific binding properties of proteins will in the end provide the key to these phenomena.

.   .   .

If one analyzes the catalytic or regulatory or epigenetic functions of proteins, one is led to the recognition that each and every one depends — above all — upon the capacities of these molecules for stereospecific association.

According to the thesis expounded in this and the two preceding chapters, all the teleonomic performances and structures of living beings are, at least in principle, analyzable in these terms. Assuming this concept to be adequate — and there is no reason to doubt that it is — the remaining step toward resolving the paradox of teleonomy is to give an explicit account of the manner in which stereospecific associative protein structures form and of the mechanisms by which they evolve. For the moment I shall focus upon their manner of forming, setting aside the question of their evolution for later chapters. I hope to show that the detailed analysis of these molecular structures, in which the ultimate "secret" of teleonomy lies hidden, leads to profoundly significant conclusions.

primary and globular structures of proteins

To begin with we must point out that the three-dimensional structure of a globular protein is determined by two types of chemical bonds.*

1. The so-called "primary" structure is constituted by a topologically linear sequence of amino acid residues linked by covalent bonds. Thus by themselves these bonds define a fibrous, exceedingly flexible structure, able in theory to take on an all but infinite variety of shapes.

---

* See Appendix 1, p. 183.

2. But the so-called "native" shape of a globular protein is in addition stabilized by a very great number of noncovalent interactions which bind together the amino acid residues distributed along the topologically linear covalent sequence. As a result the polypeptide fiber folds in a very complex way into a compact, pseudo-globular bundle. These complex foldings are what actually determine the molecule's three-dimensional structure, including the exact shape of the stereospecific binding sites by which the molecule performs its recognition activity. And so one sees that it is the sum or rather the cooperation of a multitude of noncovalent intramolecular interactions that stabilizes the functional structure of the protein — which in turn enables it to form — electively — stereospecific complexes (likewise noncovalent) with other molecules.

The question concerning us here is the ontogenesis, the origin and development of this special, this unique conformation to which a protein's cognitive function is tied. For a long time it was thought that because of the very complexity of these structures and the fact they are stabilized by noncovalent and individually very labile interactions, a vast number of different shapes would be available to a given polypeptide fiber. But an entire body of findings was to show that a given chemical species (defined by its primary structure) exists in its native state, under normal physiological conditions, only in a single conformation (or at the very most in a small number of discrete states, not very different from each other, as is the case with allosteric proteins). This is a very precisely defined conformation, as proven by the fact that protein crystals yield fine X-ray diffraction images — which means that the position of the great majority of the thousands of atoms composing a molecule is defined to within a fraction of an angstrom. We may add that this combined uniformity and precision of structure is indispensable to specific binding, a biologically essential property of globular proteins.

The mechanism whereby these structures form is today well enough understood. We are able to say that

formation of globular structures

a. The genetic determinism of protein structures *exclusively specifies the sequence* of the amino acid residues corresponding to a given protein; and

b. The polypeptide fiber thus synthesized folds in upon itself *spontaneously* and *autonomously*, ending up in its pseudo-globular, functional shape.

Thus, among the thousands of different ways in which the polypeptide fiber could theoretically bundle itself, only one is actually adopted. Here we have manifestly a true epigenetic process at the simplest possible level, that of an isolated macromolecule. To the unfolded fiber any number of conformations are open. Moreover, prior to folding it is devoid of any biological activity, and notably of any capacity for stereospecific recognition. For the folded form, on the other hand, a single shape and state actually obtains, which consequently corresponds to a much higher degree of order. With this state and with no other its functional activity is connected.

The explanation of this little miracle of molecular epigenesis is, in principle at least, relatively simple.

1. In the physiologically normal medium, i.e., in aqueous phase, the bundled state of the protein is thermo-

dynamically more stable than the unfolded one. The reason for this gain in stability is most interesting and worth noting. About one half of the amino acid residues making up the sequence are "hydrophobic," that is, they behave like oil in water: they tend to collect, freeing the water molecules immobilized through contact with them. As a result the protein assumes a compact structure, by reciprocal contact immobilizing the residues composing the fiber. Whence, for the protein, a heightening of order (negentropy)—counterbalanced within the system by an attenuation of order (i.e., an increase of entropy) caused by the admixture of the released water molecules.

2. Among the many different folded shapes accessible to a given polypeptide sequence only a very few, if not just one, will permit realization of the most compact possible structure. This structure will therefore be favored over all others. Simplifying a little, we may say that the "chosen" structure will be the one corresponding to the expulsion of the maximum number of water molecules. Clearly it is upon the relative position—that is, the sequence—of the amino acid residues in the fiber (beginning with the hydrophobic residues) that the various possibilities of achieving compact structure will depend. The globular shape peculiar to a given protein—the special shape required for its functional activity—will therefore be in fact *dictated* by the sequence of residues in the fiber. However, and this is the important point, the quantity of information that would be needed to describe the entire three-dimensional structure of a protein is *far greater* than the amount of information defined by the sequence itself. For example, for a polypeptide 100 residues long the information (H) necessary to define the sequence

would come to about 2000 bits (H = $\log_2 20^{100}$), whereas to define its three-dimensional structure this sum of information would have to be supplemented by a great deal more, the exact amount being difficult to calculate.

Thus there is a seeming contradiction between the statement that the genome "entirely defines" the function

the false paradox of genetic enrichment

of a protein and the fact that this function is linked to a three-dimensional structure whose data content is *richer* than the direct contribution made to the structure by the genome. Certain critics of modern biological theory have seized upon this contradiction, in particular Elsässer, who in the epigenetic development of the (macroscopic) structures of living beings likes to see a phenomenon beyond physical explanation, by reason of the "uncaused enrichment" it appears to indicate.

A careful and detailed scrutiny of the mechanisms of molecular epigenesis disposes of this objection. The enrichment of information evidenced in the forming of three-dimensional protein structures comes from the fact that genetic information (represented by the sequence) is expressed under strictly defined initial conditions (aqueous phase, narrow latitude of temperatures, ionic composition, etc.). The result is that of all the structures possible only one is actually realized. Initial conditions hence enter among the items of information finally enclosed within the globular structure. Without specifying it, they contribute to the realization of a unique shape by eliminating all alternative structures, in this way proposing—or rather, imposing—an unequivocal interpretation of a potentially equivocal message.

Thus the structuring process of a globular protein may

at the same time be seen as the microscopic image and as the source of the autonomous epigenetic development of the organism itself. A development in which several ascending stages or levels are discernible:

1. Folding of the polypeptide sequences culminating in globular structures provided with stereospecific binding properties.

2. Associative interactions between proteins (or between proteins and certain other constituents) so as to build cellular organelles.

3. Interactions between cells, so as to constitute tissues and organs.

4. Throughout the process, coordination and differentiation of chemical activities via allosteric-type interactions.

At each stage more highly ordered structures and new functions appear which, resulting from spontaneous interactions between products of the preceding stages, reveal successively, like a blossoming firework, the latent potentialities of previous levels. The determining cause of the entire phenomenon, its source, is finally the genetic information represented by the sum of the polypeptide sequences, interpreted—or, to be more exact, screened—by the initial conditions.

The *ultima ratio* of all the teleonomic structures and performances of living beings is thus enclosed in the sequences of residues making up polypeptide fibers, "embryos" of the globular proteins which in biology play the role Maxwell assigned to his demons a hundred years ago. In a sense, a very real sense, it is at this level of chemical organization that the secret of life lies, if indeed there is any one such secret. And if one were able not only to describe these sequences but to pronounce the law by which they assemble, one could declare the secret penetrated, the *ultima ratio* discovered.

The first description of a globular protein's complete sequence was given by Sanger in 1952. It was both a revelation and a disappointment. This sequence, which one knew to define the structure, hence the elective properties of a functional protein (insulin), proved to be without any regularity, any special feature, any restrictive characteristic. Even so the hope remained that, with the gradual accumulation of other such findings, a few general laws of assembly as well as certain functional correlations would finally come to light. Today our information extends to hundreds of sequences corresponding to various proteins extracted from all sorts of organisms. From the work on these sequences, and after systematically comparing them with the help of modern means of analysis and computing, we are now in a position to deduce the general law: it is that of chance. To be more specific: these structures are "random" in the precise sense that, were we to know the exact order of 199 residues in a protein containing 200, it would be impossible to formulate any rule, theoretical or empirical, enabling us to predict the nature of the one residue not yet identified by analysis.

To say that in a polypeptide the amino acid sequence is "random" may perhaps sound like a roundabout admission of ignorance. Quite to the contrary, the statement expresses the nature of the facts. For example: the average frequency with which such a residue in the polypeptide chain is followed by such an other is equal to the *product* of the average frequencies of each of the two res-

the *ultima ratio*
of teleonomic
structures

idues in proteins at large. We can illustrate this in another way. Imagine a deck of two hundred cards, each card marked with the name of an amino acid. A deck, moreover, in which the *average* proportion of each of the twenty amino acids would be respected. After shuffling, the cards are turned up one by one: the order in which they now appear defines a sequence *which could not be distinguished from a natural one by any objective criterion.*

But while in this sense every primary protein structure looks like the product of a random choosing from among the twenty available residues, we must recognize on the other hand that in another, equally significant sense a given sequence—*this actual one we are dealing with*—has not been synthesized at random: for the very same order it contains is reproduced, practically without error, in all the molecules of the protein under consideration. Were it not so it would be impossible, indeed, to establish the sequence of a population of molecules by chemical analysis.

And so it must be acknowledged that the "random" sequence of each protein is in fact reproduced thousands and thousands of times over, in each organism, each cell, with each generation, by a highly accurate mechanism which guarantees the invariance of the structure.

Not only the principle but most of the components of this mechanism are known today. We shall return to it in a later chapter. No detailed knowledge of its workings is needed to grasp the deep significance of the mysterious message constituted by the sequence of residues in a polypeptide fiber. A

the interpretation of the message

message which, by no matter what criteria one judges it, seems to have been composed utterly at haphazard; a message nevertheless laden with a meaning which comes out in the discriminative, functional, directly teleonomic interactions of the globular structure: the three-dimensional translation of the linear sequence. With the globular protein we already have, at the molecular level, a veritable machine—a machine in its functional properties, but not, we now see, in its fundamental structure, where nothing but the play of blind combinations can be discerned. Randomness caught on the wing, preserved, reproduced by the machinery of invariance and thus converted into order, rule, necessity. A *totally* blind process can by definition lead to anything; it can even lead to vision itself. In the ontogenesis of a functional protein are reflected the origin and descent of the whole biosphere. And the ultimate source of the project that living beings represent, pursue, and accomplish is revealed in this message—in this neat, exact, but essentially indecipherable text that primary structure constitutes. Indecipherable, since before expressing the physiologically necessary function which it performs spontaneously, in its basic make-up it discloses nothing other than the pure randomness of its origin. But such, precisely, is the profounder meaning of this message which comes to us from the most distant reaches of time.

VI
Invariance and Perturbations

EVER SINCE its birth in the Ionian Islands almost three thousand years ago, Western philosophy has been divided between two seemingly opposed attitudes. According to one of them the authentic and ultimate truth of the world can reside only in perfectly immutable forms, by essence unvarying. According to the other, the only real truth resides in flux and evolution. From Plato to Whitehead and from Heraclitus to Hegel and Marx, it is clear that these metaphysical epistemologies were always closely bound up with their authors' ethical and political biases. These ideological edifices, represented as self-evident to reason, were actually *a posteriori* constructions designed to justify preconceived ethico-political theories.*

Plato and Heraclitus

For science the only *a priori* is the postulate of objectivity, which spares—or rather forbids—it from taking part in the debate. Science studies evolution, whether that of the universe or of the systems it contains, such as the

---

* Cf. Karl Popper, *The Open Society and Its Enemies* (London: Routledge & Kegan Paul, 1945).

biosphere, including man. We are aware that any phenomenon, any event, any cognition implies interactions which by themselves generate modifications in the elements of the system. From this it does not follow that the existence of immutable entities within the structure of the universe must be denied. Quite the contrary: for the basic strategy of science in the analysis of phenomena is the ferreting out of invariants. Every law of physics, for that matter like every mathematical development, specifies some invariant relation; science's fundamental statements are expressed as universal "conservation principles." It is readily seen, by whatever example one may wish to choose, that it is in fact impossible to analyze any phenomenon otherwise than in terms of the invariants that are conserved through it. Perhaps the clearest instance of this is the formulation of the laws of kinetics, which *demanded* the invention of differential equations, that is, a means for defining change in terms of what remains unchanged.

It may be asked, of course, whether all the invariants, conservations, and symmetries that make up the texture of scientific discourse are not fictions substituted for reality in order to obtain a workable image of it—an image partially emptied of substance, but accessible to the operations of a logic itself founded upon a purely abstract, perhaps "conventional" principle of identity—a convention which, however, human reason seems incapable of doing without.

It is a classic problem, and I allude to it here in order to note that its status has been profoundly altered by the "quantic revolution." The principle of identity does not belong, as a postulate, in classical physics. There it is employed only as a logical device, nothing requiring that it be taken to correspond to a substantial reality. It is an altogether different matter in modern physics, one of whose root assumptions is the *absolute* identity of two atoms found in the same quantum state.* Whence also the absolute, nonperfectible representational value quantum theory assigns to atomic and molecular symmetries. And so today it seems that the principle of identity can no longer be confined to the status simply of a rule of logical derivation: it must be accepted as expressing, at least on the quantic scale, a substantial reality.

Be that as it may, in science there is and will remain a Platonic element which could not be taken away without ruining it. Amidst the infinite diversity of singular phenomena, science can only look for invariants.

There was a "Platonic" ambition in the systematic search for anatomical invariants to which after Cuvier—and Goethe—the great nineteenth-century naturalists devoted themselves. Modern biologists sometimes do less than full justice to the genius of the men who, behind the bewildering variety of morphologies and modes of life of living beings, succeeded in identifying, if not a unique "form," at least a finite number of anatomical archetypes, each of them invariant within the group it characterizes. It required little perceptiveness, doubtless, to see that seals are mammals and near relatives to carnivores that live on

anatomical invariants

---

* V. Weisskopf, in "Symmetry and function in biological systems at the macromolecular level," *Nobel Symposium* No. 11, ed. Engström and Strandberg, New York (1969), p. 28.

land. But it required a great deal to discern the same fundamental scheme in the anatomy of tunicates and vertebrates, permitting their joint inclusion in the phylum of Chordata. And it was still more of a feat to perceive the affinities between Chordata and echinoderms; yet it is certain, and biochemistry confirms it, that sea urchins are closer kin to us than are the members of certain far more evolved groups of invertebrates such as the cephalopods, for example.

From this immense research into basic anatomical types classical zoology and paleontology were built, a monument whose structure at once invites and justifies the theory of evolution.

The diversity of types remained even so, and there was no getting round the fact that a great many macroscopic structural patterns, radically unlike one another, coexisted in the biosphere. A blue alga, an infusorian, an octopus, and a human being – what had they in common? With the discovery of the cell and the advent of cellular theory a new unity could be seen under this diversity. But it was some time before advances in biochemistry, mainly during the second quarter of this century, revealed the profound and strict oneness, on the microscopic level, of the whole of the living world. Today we know that from the bacterium to man the chemical machinery is essentially the same, in both its structure and its functioning.

1. In its structure: all living beings, without exception,

the chemical invariants

are made up of the same two principal classes of macromolecular components: proteins and nucleic acids. What is more, these macromolecules are in all living beings constituted by the assem-

102

bling of the same residues, finite in number: twenty amino acids for the proteins and four kinds of nucleotides for the nucleic acids.

2. In its functioning: the same reactions, or rather sequences of reactions, are used in all organisms for the essential chemical operations: the mobilization and storing of chemical potential, the biosynthesis of cellular components.

True, upon this central theme of metabolism many variations are to be met with, each corresponding to a particular functional adaptation. However, they almost always consist in new utilizations of universal metabolic sequences, hitherto employed for other functions. For instance, the excretion of nitrogen occurs in different forms in birds and mammals: the former excrete uric acid, the latter urea. Now the pathway for the synthesis of uric acid in birds is only a modification, a minor one moreover, of the sequence of reactions which in all organisms synthesizes the so-called purine nucleotides, universal components of nucleic acids. In mammals the synthesis of urea is obtained thanks to a modification of another universal metabolic pathway: the one which concludes with the synthesis of arginine, an amino acid present in all proteins. Any number of examples could be given.

To biologists of my generation fell the discovery of the virtual identity of cellular chemistry throughout the entire biosphere. By 1950 research pointed to it as a certainty, and each new publication added further confirmation of it. The hopes of the most convinced "Platonists" were being more than gratified.

But this gradual disclosure of the universal "form" of cellular chemistry seemed, in the meantime, to render the problem of reproductive invariance still more acute and more paradoxical. If, chemically, the components are the same and are synthesized by the same processes in all living beings, what is the source of their prodigious morphological and physiological diversity? And, yet more puzzling, how does each species, using the same materials and the same chemical transformations as all the others, maintain, unchanged from generation to generation, the structural standard that characterizes it and differentiates it from every other?

We now have the solution to this problem. The universal components — the nucleotides on the one side, the amino acids on the other — are the logical equivalents of an alphabet in which the structure and consequently the specific associative functions of proteins are spelled out. In this alphabet can therefore be written all the diversity of structures and performances the biosphere contains. More, with each succeeding cellular generation it is the *ne varietur* reproduction of the text, written in the form of DNA nucleotide sequences, that guarantees the invariance of the species.

The fundamental biological invariant is DNA. That is why Mendel's defining of the gene as the unvarying bearer of hereditary traits, its chemical identification by Avery (confirmed by Hershey), and the elucidation by Watson and Crick of the structural basis of its replicative invariance, without any doubt constitute the most important discoveries ever made in biology. To which of course must be added the theory of natural selection, whose certainty and full significance were established only by those later discoveries.

DNA as the fundamental invariant

Invariance and Perturbations

The structure of DNA; how that structure accounts for its capacity to dictate an exact copy of the nucleotide sequence which specifies a gene; the chemical machinery that translates the nucleotide sequence of a DNA segment into an amino acid sequence in a protein—all these facts and concepts have been thoroughly and well presented for nonspecialists. No detailed review of them need be given here.* The following diagram, which sketches only in outline the two processes of *replication* and of *translation*, will suffice as basis for the present discussion:

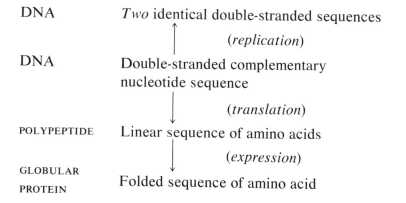

The first point that should be brought out is that the "secret" of DNA's *ne varietur* replication resides in the *stereochemical complementarity* of the *noncovalent* complex constituted by the two strands associated in the molecule. Thus we observe that the fundamental principle of associative stereospecificity, which accounts for the discriminative properties of proteins, is also at the basis of

---

* See Appendix 3, p. 189.

the replicative properties of DNA. But in DNA the complex's topological structure is far simpler than in protein complexes, and it is this that enables the replication mechanism to work. Actually, the stereochemical structure of one of the two strands is entirely defined by the sequence (the succession) of the residues composing it, because *each* of the four residues is *individually* pairable (owing to steric restrictions) with but *one* of the three others. As a result:

1. The steric structure of the complex can be completely represented in *two dimensions*, one of which, finite, contains at each point a pair of mutually complementary nucleotides, while the other contains a potentially infinite sequence of these pairs.

2. Given one – either one – of the two strands, the complementary sequence can be reconstituted step by step by successive additions of nucleotides, each of these being "chosen" by its sterically predestined partner. So it is that each of the two strands dictates the structure of its complement, so as to reconstitute the entire complex.

The DNA molecule's overall structure is the simplest and likeliest that a macromolecule constituted by the linear polymerization of identical or similar residues could adopt: that of a helix defined by two operations of symmetry, a translation and a rotation. Owing to the regularity of its structure as a whole, the DNA helix may be regarded as a fibrillar crystal. But if one considers its finer structure, it ought rather to be called an *aperiodic* crystal, since the sequence of the base pairs is nonrepetitive. It should be underscored that the sequence is entirely "free," inasmuch as no restriction is imposed upon it by the overall structure, which can accommodate all possible sequences.

106

Invariance and Perturbations

As we have just seen, the forming of this structure compares very closely with that of a crystal. Each sequential element in one of the two strands acts the part of a crystalline seed which chooses and orients the molecules that spontaneously link themselves to it, ensuring the crystal's growth. If artificially separated, two complementary strands will *spontaneously* reform the specific complex, each of them almost unerringly choosing its partner from among thousands or millions of other sequences.

However, the growth of each strand implies the formation of *covalent* bonds which sequentially interconnect the nucleotides. The formation of these bonds cannot take place spontaneously: a source of chemical potential and a catalyst are needed. The source of potential is represented by certain bonds present in the nucleotides themselves, which are split in the course of the condensation reaction. The latter is catalyzed by an enzyme, DNA polymerase. The sequence, specified by the preexisting strand, is unaffected by this enzyme. It has been shown, furthermore, that the condensation of mononucleotides activated by nonenzymatic catalyzers is actually directed by their spontaneous pairing with a preexisting polynucleotide.* Yet it is certain that while the enzyme does not specify the sequence, it does contribute to the precision of the complementary copy – that is, to the fidelity of the transfer of information. As borne out by experiment, it is an extremely high degree of fidelity; but, the process being microscopic, its accuracy cannot be absolute. This is a major point, and we shall return to it shortly.

The mechanism whereby the nucleotide sequence is

---

* L. Orgel, *Journal of Molecular Biology, 38* (1968), 381-93.

translated into an amino acid sequence is a great deal more

**the translation
of the code**

complicated even in its principle than that of *replication*. Basically, the latter process is to be explained, as we have just seen, by *direct* stereospecific interactions between a polynucleotide sequence serving as a template and the nucleotides that bind thereto. In translation non-covalent stereospecific interactions once again guarantee the transfer of information. But these governing interactions contain several successive steps, bringing into play several components each of which recognizes exclusively its immediate functional partners. The components involved at the beginning of this chain of information transfer enact their role in complete ignorance of what is "going on" at the other end of the chain. Thus, while it is true that the genetic code is written in a stereochemical language, each of whose letters consists of a sequence of three nucleotides (a triplet) in the DNA, specifying one amino acid (among twenty) in the polypeptide, there exists no direct steric relationship between the coding triplet and the coded amino acid.

Whence a most important conclusion: this code, universal in the biosphere, seems to be chemically *arbitrary*, inasmuch as the transfer of information could just as well take place according to some other convention.* Indeed, mutations are known which, impairing the structure of certain components of the translation mechanism, thereby modify the interpretation of certain triplets and thus (with regard to the convention in force) commit errors which

---

* We shall return to this point in Chapter VIII.

are exceedingly prejudicial to the organism.

The highly mechanical and even "technological" aspect of the translation process merits attention. The successive interactions of the various components intervening at each stage, leading to the assembly, residue by residue, of a polypeptide upon the surface of the ribosome, like a milling machine which notch by notch moves a piece of work through to completion – all this inevitably recalls an assembly line in a machine factory.

All told, in the normal organism this microscopic precision machinery confers a remarkable accuracy upon the process of translation. To be sure, mistakes do happen, but so rarely that no usable statistics on their normal average frequency are available. The code being unambiguous (for the translation of DNA into proteins), it follows that the sequence of nucleotides in a DNA segment entirely defines the sequence of amino acids in the corresponding polypeptide. Since, as we saw in Chapter V, the polypeptide sequence specifies completely (under normal initial conditions) the folded structure that the polypeptide adopts once it is constituted, the structural and hence functional "interpretation" of genetic information is unequivocal, rigorous. No supplementary input of information other than the genetic is necessary; none, it seems, is even possible, the mechanism as we know it leaving no room for any. And to the extent that all the structures and performances of organisms result from the structures and activities of the proteins composing them, one must regard the total organism as the ultimate epigenetic expression of the genetic message itself.

A further point should be made, again of capital importance: *the translation mechanism is strictly irreversible.*

the
irreversibility
of translation

Information is never seen being conveyed in the opposite direction—i.e., from protein to DNA—nor is there any conceivable way in which it could be. This certitude rests upon an accumulation of observations by now so complete and so well verified—and its consequences, especially for evolutionary theory, are so important—that it may be considered one of the fundamental tenets of modern biology*. Of this the upshot is that there is no *possible* mechanism whereby the structure and performance of a protein could be modified, and these modifications transmitted even partially to posterity, except by an alteration of the instructions represented by a segment of DNA sequence. Conversely, there exists no conceivable mechanism whereby any instruction or piece of information could be transferred to DNA.

Hence the entire system is totally, intensely conservative, locked into itself, utterly impervious to any "hints" from the outside world. Through its properties, by the microscopic clockwork function that establishes between DNA and protein, as between organism and medium, an entirely one-way relationship, this system obviously defies any "dialectical" description. It is not Hegelian at all,

---

* Some critics of the French edition of the present book (Piaget for instance) seemed very happy to be able to point to very recent observations as invalidating (so they thought) this statement. This claim rested on the discovery by Temin and by Baltimore of enzymes able to transcribe RNA into DNA, that is, in reverse of the operation of the more usual, already classical, systems. This important observation actually in no way violates the principle that the translation of sequential information from DNA (or from RNA) to protein is irreversible. The authors of the discovery (who are highly competent molecular biologists) did not, of course, make any such claim.

110

but thoroughly Cartesian: the cell is indeed a *machine*.

And so it would seem that by virtue of its very structure this system ought to resist all change, all evolution. Resist them it assuredly does, and we have there the explanation* for a fact which is indeed far more paradoxical than evolution itself: namely, the prodigious stability of certain species which have been able to reproduce without appreciable modification for hundreds of millions of years.

Physics tells us however that—save at absolute zero, an inaccessible limit—no microscopic entity can fail to undergo quantum perturbations, whose accumulation within a macroscopic system will slowly but surely alter its structure.

Living beings, despite the perfection of the machinery that guarantees the faithfulness of translation, are not exempt from this law. Aging and death in pluricellular organisms is accounted for, at least in part, by the piling up of accidental errors of translation. These, in particular affecting certain components responsible for the accuracy of translation, tend to precipitate further errors which, ever more frequent, gradually and inexorably undermine the structure of those organisms.†

Nor, without violating the laws of physics, could the mechanism of replication be completely immune to disturbances, accident-proof. At least some of these disturbances create more or less discrete modifications in certain elements of

microscopic perturbations

---

* The partial explanation. See p. 121.
† L. Orgel, *Proceedings of the National Academy of Science*, 49 (1963), 517.

the DNA sequence. Such errors of replication, thanks to the blind fidelity of the mechanism, will be automatically replicated again. They will be just as faithfully translated into an alteration of the amino acid sequence in the polypeptide corresponding to the DNA segment in which the *mutation* has occurred. But only when this partly new polypeptide has folded in upon itself will the functional import of the mutation become manifest.

In modern biological research some of the most outstanding work, both as to methodology and as to significance, bears upon the area known as molecular genetics (Benzer, Yanofsky, Brenner and Crick). In particular this work has made it possible to analyze the different types of discrete accidental alterations a DNA sequence may suffer. Various mutations have been identified as due to

1. The substitution of a single pair of nucleotides for another pair;

2. The deletion or addition of one or several pairs of nucleotides; and

3. Various kinds of "scrambling" of the genetic text by inversion, duplication, displacement, or fusion of more or less extended segments.*

We call these events accidental; we say that they are random occurrences. And since they constitute the *only* possible source of modifications in the genetic text, itself the *sole* repository of the organism's hereditary structures, it necessarily follows that chance *alone* is at the source of every innovation, of all creation in the biosphere. Pure chance, absolutely free but blind, at the very root of the stupendous edifice of evolution: this central

---

* See Appendix 3, p. 191.

concept of modern biology is no longer one among other possible or even conceivable hypotheses. It is today the *sole* conceivable hypothesis, the only one that squares with observed and tested fact. And nothing warrants the supposition — or the hope — that on this score our position is likely ever to be revised.

There is no scientific concept, in any of the sciences, more destructive of anthropocentrism than this one, and no other so rouses an instinctive protest from the intensely teleonomic creatures that we are. For every vitalist or animist ideology it is therefore the concept or rather the specter to be exorcized at all costs. And so it is most important to say something about the words *chance* and *randomness*, and to specify in just what sense they may and must be used with regard to mutations as the source of evolution. The idea of chance is not a simple one, and the word itself is employed in a wide variety of contexts. The best thing is to take a few examples.

Dice or roulette are termed games of chance, and the theory of probability is used to forecast their outcome. **operational uncertainty and essential uncertainty** But chance enters into these purely mechanical and macroscopic games only because of the practical impossibility of governing the throw of the dice or the spinning of the little ball with sufficient precision. An exceedingly precise mechanical thrower could conceivably be invented which would go far to reduce the uncertainty of the outcome. Let us say that in roulette the uncertainty is purely operational and not essential. The same holds, one will quickly see, for the theory of numerous phenomena where the concept of chance and the theory of probability are used for purely methodological reasons.

But in other situations the idea of chance takes on an essential and no longer merely operational meaning. Such is the case, for instance, in what may be called "absolute coincidences," those, that is to say, which result from the intersection of two totally independent chains of events. Suppose that Dr. Brown sets out on an emergency call to a new patient. In the meantime Jones the contractor's man has started making emergency repairs on the roof of a nearby building. As Dr. Brown walks past the building, Jones inadvertently lets go of his hammer, whose (deterministic) trajectory happens to intercept that of the physician, who dies of a fractured skull. We say he was a victim of chance. What other term fits such an event, by its very nature unforeseeable? Chance is obviously the essential thing here, inherent in the complete independence of two causal chains of events whose convergence produces the accident.

Now, between the occurrences that can provoke or permit an error in the *replication* of the genetic message and its functional consequences there is also a complete independence. The functional effect depends upon the structure, upon the actual role of the modified protein, upon the interactions it ensures, upon the reactions it catalyzes—all things which have nothing to do with the mutational event itself nor with its immediate or remote causes, regardless of the nature, whether deterministic or not, of those causes.

Finally, on the microscopic level there exists a further source of still more radical uncertainty, embedded in the quantum structure of matter. A mutation is in itself a microscopic event, a quantum event, to which the principle of uncertainty consequently applies. An event which is hence and by its very nature *essentially* unpredictable.

The principle of uncertainty was never entirely accepted by some of the greatest modern physicists, Einstein foremost among them, who was unwilling to admit that "God plays at dice." Certain schools have retained it for its operational usefulness but denied it the standing of an essential concept. However, all the efforts made to replace quantum theory by a "finer" structure from which uncertainty has vanished have ended in failure, and at the present time very few physicists seem disposed to believe that this principle will ever disappear from their discipline.

At any rate it must be stressed that, even were the principle of uncertainty someday abandoned, it would remain nonetheless true that between the determination, however complete, of a mutation in DNA and the determination of its functional effects on the plane of protein interaction, one could still see nothing but an "absolute coincidence" like that defined above by the parable of the workman and the physician. The occurrence would still belong to the realm of "essential" chance. Unless of course we go back to Laplace's world, from which chance is excluded by definition and where Dr. Brown has been fated to die under Jones's hammer ever since the beginning of time.

Bergson, it will be recalled, beheld in evolution the expression of an *absolutely* creative force, in the sense that he imagined it as bent on no goal other than creation in itself and for its own sake. In this he stands at the opposite pole from the animists (whether Engels,

evolution: absolute creation and not revelation

Teilhard de Chardin, or optimistic positivists like Spencer), who all regard evolution as the majestic unfolding of a program woven into the very fabric of the universe. For them, consequently, evolution is not really a creation but uniquely the "revelation" of nature's hitherto unexpressed designs. Whence the tendency to see in embryonic development an emergence of the same kind as evolutionary emergence. According to modern theory, the idea of "revelation" applies to epigenetic development, but not of course to evolutionary emergence, which, owing precisely to the fact that it arises from the essentially unforeseeable, is the creator of *absolute* newness. Would this apparent meeting of the ways between Bergsonian metaphysics and scientific thought be yet another effect of sheer coincidence? Perhaps not: artist and poet that he was, and exceedingly well acquainted with the natural sciences of his day besides, Bergson could not help being alive to the stunning richness of the biosphere and the amazing variety of forms and behavior it displays, which seem to bear almost direct witness indeed to an inexhaustible, completely untrammeled creative prodigality.

But where Bergson saw the most glaring proof that the "principle of life" is evolution itself, modern biology recognizes, instead, that all the properties of living beings rest on *a fundamental mechanism of molecular invariance.* For modern theory *evolution is not a property of living beings*, since it stems from the very *imperfections* of the conservative mechanism which indeed constitutes their unique privilege. And so one may say that the same source of fortuitous perturbations, of "noise," which in a nonliving (i.e., nonreplicative) system would lead little by little to the disintegration of all structure, is the progenitor of evolution in the biosphere and accounts for its

116

unrestricted liberty of creation, thanks to the replicative structure of DNA: that registry of chance, that tone-deaf conservatory where the noise is preserved along with the music.

VII
Evolution

THE INITIAL elementary events which open the way to evolution in the intensely conservative systems called living beings are microscopic, fortuitous, and utterly without relation to whatever may be their effects upon teleonomic functioning.

But once incorporated in the DNA structure, the accident—essentially unpredictable because always singular—will be mechanically and faithfully replicated and translated: that is to say, both multiplied and transposed into millions or billions of copies. Drawn out of the realm of pure chance, the accident enters into that of necessity, of the most implacable certainties. For natural selection operates at the macroscopic level, the level of organisms.

**chance
and necessity**

Even today a good many distinguished minds seem unable to accept or even to understand that from a source of noise natural selection alone and unaided could have drawn all the music of the biosphere. In effect natural selection operates *upon* the products of chance and can feed nowhere else; but it operates in a domain of very demanding conditions, and from this domain chance is

118

barred. It is not to chance but to these conditions that evolution owes its generally progressive course, its successive conquests, and the impression it gives of a smooth and steady unfolding.

Some post-Darwinian evolutionists have tended, when they discussed natural selection, to propagate a stark, naïvely ferocious idea of it: that of the no-holds-barred "struggle for life" – an expression which comes not from Darwin but from Herbert Spencer. The neo-Darwinians of the beginning of this century for their part proposed a much richer concept and showed, on the basis of quantitative theories, that the decisive factor in natural selection is not the struggle for life, but – within a given species – the differential rate of reproduction.

Achievements in contemporary biological research permit a sharper defining of the idea of selection. Of the intracellular cybernetic network in particular (even in the simplest organisms) – of its power, complexity, and coherence – we have established a fairly clear picture. It enables us to understand better than our less well-informed predecessors that any "novelty," in the shape of an alteration of protein structure, will be tested before all else for its compatibility with the whole of the system already bound by the innumerable controls commanding the execution of the organism's projective purpose. Hence the only acceptable mutations are those which, at the very least, do not lessen the coherence of the teleonomic apparatus, but rather, further strengthen it in its already assumed orientation or (probably more rarely) open the way to new possibilities.

It is the teleonomic apparatus, as it functions when a mutation first expresses itself, that lays down the essential *initial conditions* for the admission, temporary or permanent, or rejection of the chance-bred innovative attempt. It is teleonomic performance, the aggregate expression of the properties of the network of constructive and regulatory interactions, that is judged by selection; and that is why evolution itself seems to be fulfilling a design, seems to be carrying out a "project," that of perpetuating and amplifying some ancestral "dream."

Thanks to the conservative perfection of the replicative apparatus, any mutation, considered individually, is a very rare event. With bacteria— **the rich resources of chance** the only organisms for which we have abundant and precise data in this respect—one may say that the probability of a given gene undergoing a mutation which would significantly affect the functional properties of the corresponding protein is on the order of between one in a million and one in a hundred million per cellular generation. But a population of several billion cells can develop in a few milliliters of water. In a population of that size one may be certain that any given mutation will be represented by ten, a hundred, or a thousand samples. One may also estimate the total number of mutants of all kinds in this population at around one hundred thousand to a million.

In so large a population, consequently, mutation is by no means an exceptional phenomenon: it is the rule. And it is within the broader framework of population, not on isolated individuals, that selective pressure is exerted. The population sizes of higher organisms do not, to be sure, attain the proportions of bacterial populations; but

a. In a higher organism, for example in a mammal, the

genome contains a thousand times as many genes as the genome of a bacterium; and

b. The number of *cellular* generations, hence the number of chances for mutation, in the germinal line (i.e., the line of cells from ovule to ovule or from spermatozoon to spermatozoon) is very great in a higher organism.

This perhaps accounts for what strikes us as a relatively high incidence of certain mutations in human beings: in the vicinity of 10^{-4} to 10^{-5} for some mutations provoking easily detected genetic infirmities. It should be pointed out that the figures advanced here do not include those individually undetectable mutations which, once associated through sexual recombination, may produce significant effects. Mutations of this sort have probably had a greater importance in evolution than those whose individual effects are more pronounced.

Altogether, we may estimate that in the present-day human population of approximately three billion there occur, with each new generation, some hundred billion to a thousand billion mutations. This is only to give some idea of the extent of the vast reservoir of fortuitous variability contained within the genome of a species—again in spite of the jealously conservative properties of the replicative mechanism.

When one considers the scope of this gigantic lottery and the speed with which nature draws the numbers, one may well feel that the amazing and indeed paradoxical thing, **the "paradox" of species stability** hard to explain, is not evolution but rather the stability of the "forms" that make up the biosphere. We know that the anatomical outlines of the main phyla of the animal king-

121

dom were differentiated by the close of the Cambrian period: in other words, five hundred million years ago. It is known, besides, that certain species have remained virtually stationary for hundreds of millions of years. The lingula, for example, for the past 450,000,000 years; as for the oyster of 150,000,000 years ago, it had the appearance and probably the same flavor as those we dine on today.* Lastly, one may estimate that the present-day cell, characterized by its invariant basic chemical organization (starting with the structure of the genetic code and the complicated mechanism of translation) has been in existence for from two to three billion years, during all that time certainly provided with powerful molecular control networks guaranteeing its functional coherence.

The extraordinary stability of certain species, the billions of years spanned by evolution, the invariance of the cell's basic chemical scheme—these obviously can be explained only by the extreme coherence of the teleonomic system which in evolution has acted as both guide and brake, and has retained, amplified, and integrated only a tiny fraction of the myriad opportunities offered it by nature's roulette.

For its part the replicative system, far from being able to eliminate the microscopic perturbations by which it is inevitably beset, knows only how to register and offer them—almost always in vain—to the teleonomic filter by which their performance is finally judged, through natural selection.

---

* G. G. Simpson, *The Meaning of Evolution* (New Haven: Yale University Press, 1967).

．　　　．　　　．

A simple "point" mutation, such as the substitution of one letter in the DNA code for another, is reversible. Theory tells us that this should be so, and experiment proves it. But any appreciable evolution, like the differentiation of two even very nearly related species, reflects a great many independent mutations successively accumulated in the parent species and then, still at random, recombined thanks to the "gene flow" promoted by sexuality. Because of the number of independent events that produce it, such a phenomenon is for statistical reasons irreversible.

Evolution in the biosphere is therefore a necessarily irreversible process defining *a direction in time;* a direction which is the *same* as that enjoined by the law of increasing entropy, that is to say, the second law of thermodynamics. This is far more than a mere comparison: the second law is founded upon considerations *identical* to those which establish the irreversibility of evolution. Indeed, *it is legitimate to view the irreversibility of evolution as an expression of the second law in the biosphere.* The second law, formulating only a statistical prediction, of course does not deny any macroscopic system the possibility of facing about and, with a motion of very small amplitude and for a very brief space, reascending the slope of entropy—taking, as it were, a short step backward in time. In living beings it is precisely these fugitive stirrings which, snapped up and reproduced by the replicative mechanism, have been retained by selection. In

**irreversibility
of evolution
and the second law**

this sense natural selection – based upon a choice of rare and precious incidents contained, along with an infinity of others, within the huge reservoir of microscopic chance – constitutes a kind of Wellsian time machine.

It is not surprising but altogether natural that the results obtained by this mechanism for moving backward in time – e.g., the general upward course of evolution, the perfecting and enrichment of the teleonomic apparatus – should appear miraculous to some, paradoxical to others, and that the modern "Darwinian-molecular" theory of evolution should even today be held in suspicion by certain thinkers: philosophers or, for that matter, biologists.

This is owing, at least in part, to the extreme difficulty one has in imagining the inexhaustible resources of that ocean of chance upon which selection draws. Yet a remarkable illustration of it may be found in the organism's system of defense through antibodies. These are proteins endowed with the capacity to recognize, by stereospecific

origin of antibodies

association, substances foreign to the organism which have invaded it: for example bacteria and viruses. But as we all know, the antibody that electively recognizes a given substance – for example, a "steric pattern" peculiar to a certain bacterial species – makes its appearance in the organism (where it will remain present for some time) only after the organism has had at least one experience with the intruder (through vaccination, spontaneous or artificial). It has been further demonstrated that the organism is capable of forming antibodies equipped to cope with practically any natural or synthetic steric pattern. The possibilities, in this respect, seem virtually without limit.

124

And so for a long time it was supposed that the source of information for the synthesis of the antibody's specific associative structure was the antigen itself. Today however it is known that the structure of the antibody owes nothing to the antigen. Within the organism specialized cells, produced in great number, possess the property – the unique property – of "playing roulette" with a well-defined part of the genetic segments that determine the structure of antibodies. The exact functioning of this specialized and ultrarapid genetic roulette has not been entirely elucidated as yet: it seems likely, though, that recombinations as well as mutations occur, but in either case occur at random, in completest ignorance of the structure of the antigen. Opposite this random proliferation the antigen plays the part of selector, differentially favoring the multiplication of those cells which happen to produce an antibody capable of recognizing it.

It is indeed remarkable to find chance at the basis of one of the most exquisitely precise adaptation phenomena we know. But it is clear (after the fact) that only such a source as chance could be rich enough to supply the organism with means to repel attack from any quarter.

Other difficulties about accepting the selective theory are to be traced to its having been too often understood or represented as placing the sole responsibility for selection upon conditions of the external environment. This is a completely mistaken conception. For the selective pressures exerted by outside conditions upon organisms are in no case unconnected with the teleonomic performances characteristic of the species. Different organisms

**behavior orients
the pressures
of selection**

inhabiting the same ecological niche interact in very different and specific ways with outside conditions (among which one must include other organisms). These specific interactions, which the organism itself "elects," at least in part, determine the nature and orientation of the selective pressure the organism sustains. Let us say that the "initial conditions" of selection encountered by a new mutation simultaneously and inseparably include both the environment surrounding the organism and the total structures and performances of the teleonomic apparatus belonging to it.

Obviously, the part played by teleonomic performances in the orientation of selection becomes greater and greater, the higher the level of organization and hence *autonomy* of the organism with respect to its environment —to the point where teleonomic performance may indeed be considered decisive in the higher organisms, whose survival and reproduction depend above all upon their behavior.

It is evident as well that the initial choice of this or that kind of behavior can often have very long-range consequences, affecting not only the species in which it first crops up in rudimentary form, but all its descendants, even if these should constitute an entire evolutionary subgroup. As we all know, the great turning points in evolution have coincided with the invasion of new ecological spaces. If terrestrial vertebrates appeared and were able to initiate that wonderful line from which amphibians, reptiles, birds, and mammals later developed, it was originally because a primitive fish "chose" to do some exploring on land, where it was however ill-provided with means for getting about. The same fish thereby created, as a consequence of a shift in behavior, the selective pressure which was to engender the powerful limbs of the quadrupeds. Among the descendants of this daring explorer, this Magellan of evolution, are some that can run at speeds of fifty miles an hour; others climb trees with astonishing agility, while yet others have conquered the air, in a fantastic manner fulfilling, extending, and amplifying the ancestral fish's hankering, its "dream."

The fact that in the evolution of certain groups one observes a general tendency, maintained over millions of years, toward the apparently oriented development of certain organs shows how the initial choice of a certain kind of behavior (for example, in the face of attack from a predator) commits the species irrevocably in the direction of a continuous perfecting of the structures and performances this behavior needs for its support. It is because the ancestors of the horse at an early point chose to live upon open plains and to flee at the approach of an enemy (rather than try to put up a fight or hide) that the modern species, following a long evolution made up of many stages of reduction, today walks on the tip of a single toe.

It is known that certain very precise and complex kinds of behavior, such as the prenuptial ceremonies of birds, are narrowly linked to certain especially conspicuous morphological features. There can be no doubt that this behavior and the anatomical particularities that go with it evolved *pari passu*, each encouraging and reinforcing the other under the pressure of sexual selection. Once it starts to develop in a species, any decorative finery associated with successful mating only adds to, in short confirms, the initial pressure of selection and consequently favors any improvement in the finery itself. One is therefore quite right in saying that the sexual drive—or better still, *desire*—created the conditions under which some magnificent plumages were selected.*

Lamarck was of the belief that the very strain entailed in an animal's efforts to "succeed" in life somehow affected its hereditary legacy, entering into it and having a direct modeling influence upon its descendants. The giraffe's immensely long neck would thus express its forebears' unabating wish to reach the topmost branches of trees. This is of course today an unacceptable hypothesis; yet one sees that pure selection, operating upon elements of behavior, leads to the result Lamarck sought to explain: the close interconnection of anatomical adaptations and specific performances.

It is in these terms that one must confront the problem of the selective pressures which have oriented human evolution. An exceptionally interesting problem it is, and not just because of our involvement in it, nor because from a better insight into the evolutionary roots of our being we might perhaps gain a better understanding of its present nature. An impartial observer, let's say someone from Mars, could not fail to be struck by the fact that the development of man's specific performance, symbolic language—a unique occurrence in the biosphere—opened the way for *another* evolution, creator of a new kingdom: that of culture, of ideas, of knowledge.

---

* Cf. N. Tinbergen, *Social Behavior in Animals* (London: Methuen & Co., 1953).

A unique event: modern linguists dwell upon the fact that the symbolic language of human beings is of an utter-

language and the evolution of man

ly different order from the various (auditory, tactile, visual, and other) means of communication animals employ. This is no doubt true. But to go on from there to maintain that the phenomenon attests to an absolute break in evolutionary continuity – that human language has owed nothing whatever, even *at the very outset*, to a system of various calls and warnings like those exchanged by apes – this would seem to me a rather difficult step to take, and in any case an unnecessary hypothesis.

Animals, and not only those nearest us on the evolutionary scale, unquestionably possess a brain capable not just of retaining and recording pieces of information but also of associating and transforming them, of bringing the result of these operations back out in the form of an individual performance; yet not – and this is the essential point – in a form which permits the communication to another individual of an original, personal association or transformation. But this is what can be done with human language, which may be considered by definition to have been born on the day when creative combinations – *new* associations achieved by one person – by reason of their transmission to others no longer had to perish with him.

No primitive language exists for us to study: in all the races of our unique modern species the symbolic instrument has attained roughly the same level of complexity and communicative power. Moreover, according to Chomsky the underlying structure, the "form" of all human languages is the same. The extraordinary feats that language both represents and makes possible are obviously connected with the considerable development of the central nervous system in Homo sapiens; a development which, for that matter, constitutes his most distinctive anatomical feature.

From what we know of man's most distant ancestors, we are in a position to state that his evolution has been marked above all else by the progressive development of the skull, hence of the brain. This has required over two million years of oriented and sustained selective pressure. That pressure must have been heavy, for in evolutionary terms two million years is a relatively short span; and it was specific, for we observe nothing similar in any other line: the cranial capacity of present-day apes is hardly greater than that of their forebears of several million years ago.

Between the privileged evolution of man's central nervous system and that of the unique performance which characterizes him, one must inevitably imagine a parallel, hand-in-hand development, in which language was not only the product but one of the initial conditions of this evolution.

The likeliest hypothesis in my own view is that, appearing very early in our line, the most rudimentary symbolic communication, through the radically new possibilities it offered, constituted one of those crucial initial "choices" which are binding upon the future of a species in that they give rise to a new selective pressure. This selection must have favored the development of linguistic ability itself and hence the development of the brain, the organ that serves it. I believe there are powerful arguments for this hypothesis.

The earliest authentic hominids we know of today – the australopithecines, whom Leroi-Gourhan rightly prefers to call "australanthropes" – already possessed, as their defining characteristics, those which separate man from his nearest cousins, the Pongidae (that is to say, the anthropoid apes). The australanthropes had adopted the upright posture, associated not only with specialization of the foot but with numerous muscular and skeletal modifications, notably of the vertebral column and the position of the skull with respect to it. Except for the gibbon, every anthropoid moves on all fours; as has often been said, man's evolution must have been tremendously spurred when, standing erect, he freed his hands for other purposes than use in walking. No doubt this invention of his, a very ancient one (prior to the australanthropes), was of extreme importance: for this alone permitted our ancestors to become hunters able to use their two fore-limbs while continuing to walk or run.

The cranial capacity of these primitive hominids was, however, scarcely above that of a chimpanzee and just below a gorilla's. What the brain can perform is not proportional to its weight; but its weight does doubtless impose limits to intelligence, and Homo sapiens surely could not have emerged but for the development of his skull.

At any rate it appears established that, although the brain of Zinjanthropus weighed no more than a gorilla's, it was nonetheless capable of feats unknown among the Pongidae: Zinjanthropus manufactured tools. They were very primitive ones, it is true, and recognizable as artifacts only through the repetition of the same very crude shapes and their location in the vicinity of certain fossil remains. The larger apes utilize natural "tools"—stones or branches of trees—when occasion arises, but they produce nothing comparable to artifacts fashioned according to a recognizable *norm*.

Thus Zinjanthropus must be considered a very primitive Homo faber. Now it seems likely that there must have been a close correlation between the development of language and that of an industry betokening purposive and disciplined activity.* Hence it would be reasonable to suppose that the australanthropes possessed an instrument of symbolic communication proportionate to their rudimentary industry. Furthermore, if it is true, as Dart believes,† that the australanthropes successfully hunted such powerful and dangerous beasts as the rhinoceros, the hippopotamus, and the panther, then they must have hunted as a group carrying out a previously concerted project. Language would have been required for its preliminary formulation.

This hypothesis is not vitiated by the fact that the australanthropes' brain was of very modest size. For recent experiments with a young chimpanzee seem to show that, while apes are incapable of learning spoken language, they can assimilate and utilize some elements of the sign language deaf-mutes employ.** Hence there are grounds for supposing that the acquisition of the power of articu-

---

* Leroi-Gourhan, *Le Geste et la Parole* (Paris: Albin-Michel, 1964); R. L. Holloway, *Current Anthropology, 10* (1969), 395; J. Bronowski, in *To Honor Roman Jakobson* (Paris: Mouton, 1967).

† Cited by Leroi-Gourhan, *op. cit.*

**B. T. Gardner and R. A. Gardner, in *Behavior of Non-Human Primates*, ed. Schrier and Stolnitz (New York: Academic Press, 1970).

late symbolization might have followed upon some not necessarily very elaborate neurophysiological modifications in an animal which at this stage was no more intelligent than a present-day chimpanzee.

But it is evident that, once having made its appearance, language, however primitive, could not help but greatly increase the survival value of intelligence, and thus create, in favor of the development of the brain, a formidable and oriented selective pressure the likes of which no speechless species could ever experience. As soon as a system of symbolic communication came into being, the individuals or rather the groups best able to use it acquired over others an advantage incomparably greater than a similar superiority of intelligence would have conferred upon a species without language. We see as well that the selective pressure engendered by speech was bound to steer the evolution of the central nervous system in the direction of a special kind of intelligence: the kind most apt to exploit this particular, specific performance, rich in immense possibilities.

Attractive and reasonable, this hypothesis would have little else in its favor were it not warranted by certain linguistic evidence being compiled today. The study of children's acquisition of language suggests in the most

the primary acquisition of language

compelling manner that this so astonishing process is by its very nature profoundly different from the orderly apprenticeship of a system of formal rules.* The

---

* E. Lenneberg, *Biological Foundations of Language* (New York: John Wiley & Sons, 1967).

child learns no rules, and he does not seek to imitate adult speech. Rather, one might say that he takes from it whatever suits him at each stage of his linguistic development. At the very beginning (anywhere between fourteen and eighteen months) the child disposes of a stock of maybe ten words, which he uses separately, without ever associating them even by imitation. Later he will combine words two, three, and more at a time, according to a syntax which once again is not a mere repetition or copying of what he hears from grownups. This process, it appears, is universal and its chronology the same for all tongues. The ease with which, in two or three years (after the first year), this playing with language brings the child to mastery of it is something the adult observer is always hard put to believe.

All this, one is driven to assume, must reflect an embryological, an epigenetic process in the course of which the neural structures that underlie linguistic performances develop. This assumption is borne out by observations gathered upon trauma-provoked aphasias. The younger the child in whom these aphasias occur, the more quickly and completely they tend to regress. But the impairment becomes irreversible if the lesions are suffered at the approach of puberty or later. An entire body of findings, over and above these, confirm that there exists a critical age for the spontaneous acquisition of language. As everybody knows, for an adult to learn a second language demands a great deal of determined and systematic effort; and the status of the language thus learned practically always remains inferior to that of the native tongue, spontaneously acquired.

Anatomical evidence confirms the idea that the primary

134

acquisition of language is bound up with a process of epigenetic development. It is known that the maturing of the brain, continuing after birth, halts with puberty. This development seems to consist mainly in a considerable amplification of the network of interconnections between cortical neurons. Very rapid during the first two years, the process afterward slows down.

the acquisition of language programmed in the epigenetic development of the brain

Judging from anatomical evidence it does not extend beyond puberty; it therefore coincides with the "critical period" during which primary acquisition is possible.*

On the basis of the foregoing one may venture—as, for my own part, I am quite prepared to do—that if, in the child, the acquisition of language appears so miraculously spontaneous it is because it forms part and parcel of an epigenetic development *one of whose functions is to make way for language.* To be a little more precise: the development of the cognitive function itself depends, beyond any doubt, upon this postnatal growth of the cortex. It is the acquisition of language in the very course of this epigenesis that makes for its association with the cognitive function—an association so intimate that we find it exceedingly difficult to separate, by introspection, the utterance from the thought it expounds.

Language, according to the prevailing view, is no more than a "superstructure," and this is indeed what it would seem in the light of the great diversity of human languages, products of the second evolution, that of culture.

---

* *Ibid.*

However, the extensiveness and refinement of the cognitive functions in Homo sapiens clearly find their *raison d'être* in and through language alone. Deprived of this instrument, they are for the most part rendered useless, paralyzed. Thus approached, the capacity for language can no longer be regarded as a superstructure. Rather it must be conceded that, between the cognitive functions and the symbolic language they beget—and through which they are articulated—there is in modern man a close symbiosis which can only be the product of a common evolution begun long ago.

According to Chomsky and his school, in-depth linguistic analysis reveals, beneath their boundless diversity, one basic "form" common to all human languages. Therefore Chomsky feels this form must be considered *innate* and characteristic of the species. Certain philosophers or anthropologists have been scandalized by this thesis, in it discerning a return to Cartesian metaphysics. Provided its implicit biological content be accepted I see nothing wrong with it whatsoever. On the contrary, it strikes me as a most natural conclusion, once one assumes that the evolution of man's cortical structures could not help but be extensively influenced by a capacity for language acquired very early and in the crudest possible state. Which amounts to assuming that spoken language, when it appeared among primitive mankind, not only made possible the evolution of culture but contributed decisively to man's *physical* evolution. If these are correct assumptions, the linguistic capacity that declares itself in the course of the brain's epigenetic development is today part of "human nature," itself defined within the genome in the radically different language of the genetic code. A miracle? To be sure, since in the final analysis language too was a product of chance. But on the day Zinjanthropus or one of his comrades first used an articulate symbol as representing a category, he enormously increased the probability that at some later day a brain might emerge capable of conceiving the Darwinian theory of evolution.

VIII
The Frontiers

WHEN ONE ponders on the tremendous journey of evolution over the past three billion years or so, the prodigious

the present
frontiers
of knowledge
in the field
of biology

wealth of structures it has engendered, and the extraordinarily effective teleonomic performances of living beings, from bacteria to man, one may well find oneself beginning to doubt again whether all this could conceivably be the product of an enormous lottery presided over by natural selection, blindly picking the rare winners from among numbers drawn at utter random.

While one's conviction may be restored by a detailed review of the accumulated modern evidence that this conception alone is compatible with the facts (notably with the molecular mechanisms of replication, mutation, and translation), it affords no synthetic, intuitive, and immediate grasp of the vast sweep of evolution. The miracle stands "explained"; it does not strike us as any less miraculous. As François Mauriac wrote, "What this professor says is far more incredible than what we poor Christians believe."

The Frontiers

This is true, just as it is true that there is no achieving a satisfactory mental image of certain abstractions in modern physics. But we also know that such difficulties cannot be taken as arguments against a theory which is vouched for by experiment and logic. In the case of physics, microscopic or cosmological, we see right away what the trouble is: the scale of the envisaged phenomena transcends the categories of our immediate experience. Only abstraction can supply this deficiency, yet without curing it. Where biology is concerned the difficulty is of another order. The elementary interactions upon which everything hinges, these, thanks to their "mechanical" character, are relatively easy to grasp. Much less readily come by is an intuitive global picture of living systems whose phenomenal complexity defies assimilation. But in biology as in physics these psychological difficulties, once again, do not constitute an argument against theory and observation.

Today we can assert that the elementary mechanisms of evolution have been not only understood in principle but identified with precision. The solution found is the more satisfactory since the mechanisms involved are the same ones that ensure the stability of species: replicative invariance in DNA and teleonomic coherence in organisms.

Central to biology, the evolutionary concept is bound to undergo considerable elaboration in the years ahead. Much remains to be learned. Essentially, however, the problem has been resolved and evolution now lies well to this side of the frontier of knowledge.

The present challenge, as I see it, is in the areas at the two extremes of evolution: the origin of the first living systems on the one hand; on the other, the inner workings

139

of the most intensely teleonomic system ever to have emerged, to wit, the central nervous system of Man. In this chapter I should like to try to delimit these two borderlands of the unknown.

The discovery of the universal mechanisms basic to the essential properties of living beings ought, one would think, to have facilitated solving the problem of life's origins. As it turns out these discoveries, by almost entirely transforming the question (today posed in much more precise terms) have revealed it to be even more difficult than it formerly appeared.

Three presumptive stages in the process which led to the emergence of the first organisms may *a priori* be distinguished:

the problem of life's origins

1. The formation upon earth of the main chemical building blocks of living beings: nucleotides and amino acids.

2. The formation from these materials of the first macromolecules capable of replication.

3. The evolution which elaborated a teleonomic apparatus around these "replicative structures," eventually leading to the primitive cell.

Different problems arise in the interpretation of each of these stages. The first of them, often called the "prebiotic" phase, is open enough to theoretical and indeed to experimental study. While uncertainty remains, and will doubtless continue, as to the paths prebiotic chemical evolution actually followed, the overall picture seems fairly clear. Four billion years ago atmospheric conditions and those prevailing upon the earth's surface favored the accumulation of certain simple carbon compounds such as methane. There was also water and ammonia. Now from these simple compounds and in the presence of nonbiological catalysts it is fairly easy to obtain numerous more complex compounds, among which figure some amino acids and some precursors of nucleotides (nitrogenous bases, sugars). Remarkably enough, under certain altogether plausible sets of conditions, these syntheses yield a very high percentage of compounds identical or analogous to those that enter into the makeup of the modern cell.

And so it may be considered as *proven* that at a given moment in the earth's history certain bodies of water *could* have contained in solution high concentrations of the essential components of the two classes of biological macromolecules, nucleic acids and proteins. In this prebiotic "soup" various macromolecules might have formed through polymerization of their precursors, amino acids and nucleotides. In the laboratory, under "plausible" conditions, some polypeptides and polynucleotides similar in general structure to "modern" macromolecules have actually been obtained.

Hence no major difficulties up to this point. But the decisive step has yet to be taken, from the first stage to the second: the formation of macromolecules capable, under the conditions prevailing in the primordial soup, of promoting their own replication unaided by any teleonomic apparatus. This difficulty does not seem insurmountable. It has been demonstrated that a polynucleotide sequence is effectively able to guide, by spontaneous base pairing, the synthesis of the complementary sequence. To be sure, such a mechanism could only have been very inefficient and subject to innumerable errors. But the moment it got under way, the three fundamental processes — replication, mutation, and selection — were at work and must have bestowed a heavy advantage upon the macromolecules most able, by their sequential structure, to replicate spontaneously.*

The third step, according to our hypothesis, was the gradual emergence of teleonomic systems which, around replicative structures, were to construct an *organism*, a primitive cell. It is here that one reaches the real "sound wall," for we have no idea what the structure of a primitive cell might have been. The simplest living system known to us, the bacterial cell, a tiny piece of extremely complex and efficient machinery, attained its present state of perfection perhaps a billion years ago. Its overall chemical ground plan is the same as that of all other living beings. It employs the same genetic code and the same mechanism of translation as do, for example, human cells.

Thus the simplest cells available to us for study have nothing "primitive" about them. Selection operating over five hundred or a thousand billion generations has left them with a teleonomic apparatus so powerful that no vestiges of truly primitive structures are discernible. Without the help of fossils, such an evolution cannot possibly be reconstructed. Still, one would like at least to try to suggest a plausible hypothesis as to the route this evolution followed, and especially as to the point where it began.

The development of the metabolic system, which, as

---

* L. Orgel, *Proceedings of the National Academy of Science, 49* (1963).

the primordial soup thinned, must have "learned" to mobilize chemical potential and to synthesize the cellular components, poses Herculean problems. So also does the emergence of the selectively permeable membrane without which there can be no viable cell. But the major problem is the origin of the genetic code and of its translation mechanism. Indeed, instead of a problem it ought rather to be called a riddle.

The code is meaningless unless translated. The modern cell's translating machinery consists of at least fifty macromolecular components *which are themselves coded in DNA: the code cannot be translated otherwise than by products of translation.* It is the modern expression of *omne vivum ex ovo*. When and how did this circle become closed? It is exceedingly difficult to imagine. But the fact that the code is now deciphered and known to be universal at least allows us to frame the problem in precise terms; simplifying just a little, in those of the following alternatives. Either

**the riddle of
the code's origins**

a. Chemical—or, to be more exact, stereochemical—reasons account for the structure of the code; if a certain codon was "chosen" to represent a certain amino acid it is because there existed a certain stereochemical affinity between them; or else

b. The code's structure is chemically arbitrary: the code as we know it today is the result of a series of random choices which gradually enriched it.

The first of these hypotheses seems by far the more appealing. To begin with, because it would explain the universality of the code. Next, because it permits us to imagine a primitive translation mechanism in which the sequential aligning of amino acids to form a polypeptide would be due to a direct interaction between the amino acids and the replicative structure itself. Finally and above all, because in principle this hypothesis, if true, would be verifiable. And numerous attempts to verify it have indeed been made: on the whole they have proven negative to date.*

It may be that we have yet to hear the last word on this score. Pending the not very likely confirmation of this first hypothesis we are reduced to the second, displeasing from the methodological viewpoint – which does not by any means signify that it is incorrect. Displeasing on other grounds also. It does not explain the code's universality. One is brought, then, to assume that out of a multitude of efforts at elaboration a single one survived. Which in itself makes sense, but leaves us unprovided with any model of primitive translation. Here speculation must take over. Much that is very ingenious has been put forward: the field is only too open.

The riddle remains, and in so doing masks the answer to a question of profound interest. Life appeared on earth: what, *before the event*, were the chances that this would occur? The present structure of the biosphere far from excludes the possibility that the decisive event occurred *only once*. Which would mean that its *a priori* probability was virtually zero.

This idea is distasteful to most scientists. Science can neither say nor do anything about a unique occurrence. It

---

* Cf. F. Crick, *Journal of Molecular Biology, 38* (1968), pp. 367– 79.

144

can only consider occurrences that form a class, whose *a priori* probability, however faint, is yet definite. Now through the very universality of its structures, starting with the code, the biosphere looks like the product of a unique event. It is possible of course that its uniform character was arrived at by elimination through selection of many other attempts or variants. But nothing compels this interpretation.

Among all the occurrences possible in the universe the *a priori* probability of any particular one of them verges upon zero. Yet the universe exists; particular events must take place in it, the probability of which (before the event) was infinitesimal. At the present time we have no legitimate grounds for either asserting or denying that life got off to but a single start on earth, and that, as a consequence, before it appeared its chances of occurring were next to nil.

Not only for scientific reasons do biologists recoil at this idea. It runs counter to our very human tendency to believe that behind everything real in the world stands a necessity rooted in the very beginning of things. Against this notion, this powerful feeling of destiny, we must be constantly on guard. Immanence is alien to modern science. Destiny is written concurrently with the event, not prior to it. Our own was not written before the emergence of the human species, alone in all the biosphere to utilize a logical system of symbolic communication. Another unique event, which by itself should predispose us against any anthropocentrism. If it was unique, as may perhaps have been the appearance of life itself, then before it did appear its chances of doing so were infinitely slender. The universe was not pregnant with life nor the biosphere with man. Our number came up in the Monte Carlo game. Is it any wonder if, like the person who has just made a million at the casino, we feel strange and a little unreal?

The logician might be moved to remind the biologist that his efforts to "understand" the entire functioning of the human brain are ordained to failure, since no logical system can produce an integral description of its own structure. This warning would be a little unseasonable, considering how far we still are from that ultimate borderline of knowledge. At any rate this logical objection does not apply to the analysis by man of the central nervous system of an animal. That system, we may at least suppose, would be less complex and less powerful than our own. But in this case, too, a major difficulty remains: an animal's conscious experience is and no doubt will always be impenetrable to us. So long as this is so, it is questionable whether any exhaustive description of the workings of, say, the brain of a frog is basically possible. Though encumbered by restrictions, nothing will ever be a suitable substitute for the exploration of the human brain, affording as it does the possibility of comparing objective experimental data with the facts of subjective experience.

the other frontier: the central nervous system

In any case, the structure and functioning of the brain can and must be explored simultaneously at every accessible level in the hope that these investigations, very different both in their methods and in their immediate object, will someday converge. About the only convergence they show at present is in the difficulty of the problems they all raise.

Among the knottiest and most important are the problems surrounding the epigenetic development of a structure as complex as the central nervous system. In man it contains from one to ten thousand billion neurons interconnected by means of about a hundred times that many synapses, certain of which connect nerve cells lying far apart from each other. I mentioned earlier the as yet unresolved question of how long-distance morphogenetic interactions are established. If nothing else, such problems can at least be clearly posed, thanks notably to some remarkable experimental work.*

An understanding of the central nervous system's functioning must begin with that of the synapse, its primary logical element. Investigation is easier here than at any other level, and refined techniques have yielded a considerable mass of findings. However, we are still a long way from an interpretation of synaptic transmission in terms of molecular interaction. Yet that is a most essential question, for therein probably lies the ultimate secret of memory. Quite some time ago it was proposed that the memory trace is registered in the form of a more or less irreversible alteration of the molecular interactions responsible for transmitting the nerve impulse through synapses. This theory has plausibility in its favor but no direct proofs.†

---

* R. W. Sperry, *passim*.

† A theory according to which the memory is coded in the sequences of residues of certain macromolecules (ribonucleic acids) recently found acceptance among certain physiologists. They apparently believed they had thereby linked up with and could borrow from the concepts derived from the study of the genetic code. However, this

Despite this profound ignorance concerning the fundamental mechanisms of the central nervous system, modern electrophysiology, examining the integration of nerve signals, especially in certain sensory pathways, has produced highly significant results.

First of all, with respect to the properties of the neuron as integrator of the signals it may receive, through the intermediary of synapses, from numerous other cells: analysis has shown that in its performances the neuron closely resembles the integrated components of an electronic computer. For example, like the latter it is capable of carrying out all the logical operations of propositional algebra. More, it can add or subtract different signals while taking into account their coincidence in time; it can modify the *frequency* of the signals it transmits in keeping with the *amplitude* of those it receives. In fact, it seems as though no unitary component being utilized nowadays in modern computers is capable of such varied and finely modulated performances. Between cybernetic machines and the central nervous system the analogy remains impressive and the comparison fruitful; but we must note that, at present, the parallel is confined to the lower levels of integration: the initial stages of sensory analysis, for example. The higher functions of the cortex, which achieve expression through language, seem to elude such methods of approach. One may wonder here whether the difference is "quantitative" (a greater degree of complexity) or "qualitative." This is not a meaningful question in my opinion. Nothing warrants the supposition that the basic interactions are different in nature at different levels of integration. But if a case exists where the first law of dialectics is applicable, this indeed is it.

---

theory is untenable in the light, precisely, of what we now know about the code and the mechanisms of translation.

. . .

The very refinement of the cognitive functions in man, and the copious uses he puts them to, so overshadow as to make us forget the prime functions discharged by the brain in the animal series (to which man belongs). These prime functions might be listed and defined in the following way:

functions of the central nervous system

1. To control and coordinate neuromotor activity, notably in accord with sensory inputs.

2. To contain, in the form of genetically determined elements of circuitry, more or less complex programs of action, and to set them in motion in response to particular stimuli.

3. To analyze, sift, and integrate sensory inputs so as to obtain a representation of the outside world geared to the animal's specific performances.

4. To register events which, by the yardstick of those specific performances, are significant; to group them into classes according to their analogies; to associate these classes according to the relationship (of coincidence or succession) of the events constituting them; and to enrich, refine, and diversify the innate programs by incorporating these experiences into them.

5. To imagine, that is to say, *represent* and *simulate* external events and programs of action for the animal itself.

The functions noted in the first three paragraphs are fulfilled by the central nervous system of, for example, the Arthropoda, which are not usually reckoned among the higher animals. The most spectacular examples we know of very complex innate programs of action are met with in insects. It is doubtful whether among these animals the functions outlined in paragraph (4) play an important role;* on the other hand, they contribute in a most important manner to the behavior of the higher invertebrates, such as the octopus,† and of course to that of all the vertebrates.

As for what we may term the "projective" functions in paragraph (5), they are probably the prerogative of the higher vertebrates, perhaps of mammals only. But here consciousness becomes a barrier, and it may be that we can perceive the outward signs of this activity (dreaming, for example) only in our nearer cousins, without it being totally absent in other species.

The functions cited under (4) and (5) are cognitive, while those in paragraphs (1), (2), and (3) are solely coordinative and representational. Only the functions in the last paragraph can be creative of subjective experiment.

According to the proposition in paragraph (3), the central nervous system's analysis of sense impressions furnishes a meager and slanted image of the world outside; a kind of **the analysis of sense impressions** résumé where the emphasis and focus are exclusively upon what the animal's specific behavior dictates to be of special interest to it. (An at bottom "critical" résumé, the word being taken in a complementary acceptation of the Kantian sense.) Experiment abundantly bears this out. For instance, the analyzer situated behind

---

* Save perhaps in the case of bees.

† J. Z. Young, *A Model of the Brain* (Oxford: Clarendon Press, 1964).

the eye of a frog permits it to see a fly (i.e., a black speck) that is moving, but not a fly at rest.* And so the frog will only catch its prey in flight. We must stress, and electro-physiological investigation has proven, that this behavior is not by any means indicative of the frog's disdain for a motionless black speck having none of the earmarks of food. The image of the motionless speck is surely enough registered upon the frog's retina; but it is not *transmitted*, the system being excited only by a moving object.

Certain experiments upon cats† suggest an interpreta-tion of the strange fact that a field simultaneously reflect-ing all the colors of the spectrum is seen as a *white* ex-panse, although white is subjectively beheld as complete absence of color. The experimenters have shown that due to cross inhibitions between certain neurons responding to various wavelengths, these neurons do not send signals when the retina is uniformly exposed to the entire gamut of visible wavelengths. Thus, in a subjective sense, Goethe was right in his dispute with Newton. Which is the kind of error that is eminently forgivable in a poet.

There is no doubt either that animals are able to classify objects or relationships between objects according to ab-stract categories, notably geometrical ones: an octopus or a rat can learn to distinguish such figures as a triangle, circle, or square, and recognize them unfailingly by their geometrical features, regardless of size, orientation, or the coloring of the real object presented to them.

Study of the circuits which analyze the figures placed in a cat's field of vision demonstrates that these recogni-

---

* H. B. Barlow, *Journal of Physiology, 119* (1953), 69–88.
† T. N. Wiesel and D. H. Hubel, *Journal of Neurophysiology, 29* (1966), 1115–56.

tions of geometry are owing to the structure itself of the circuits that filter and recompose the retinal image. Actually, these analyzers impose a restrictive grid upon the image, from which they extract certain simple elements. Some nerve cells, for example, respond only to the figure of a straight line sloping down from left to right; others to a line inclined in the opposite direction. Thus it is not so much that a clear geometrical "idea" is conveyed by the image of the object; rather, the sense analyzer perceives and recomposes the object out of its simplest geometrical elements.*

These contemporary discoveries therefore furnish a support—in a new sense and a different context—to Descartes and Kant and urge against the uncompromising empiricism which has held almost continual sway in science for the past two hundred years, casting suspicion upon any hypothesis positing the innateness of cognitive frames of reference. Certain contemporary ethologists still seem attached to the idea that the elements of animal behavior are some of them innate, some learned, the one mode of acquisition being strictly separate from and absolutely excluding the other. How completely mistaken this conception is has been vigorously demonstrated by Lorenz.† When behavior implies elements acquired through experience,

empiricism and innateness

---

* D. H. Hubel and T. N. Wiesel, *Journal of Physiology, 148* (1959), 574-91.

† K. Lorenz, *Evolution and Modification of Behavior* (Chicago: University of Chicago Press, 1965).

152

they are acquired according to a *program*, and that program is innate—that is to say, genetically determined. The program's structure initiates and guides early learning, which will follow a certain preestablished pattern defined in the species' genetic patrimony. Thus, in all likelihood, is the process to be understood whereby the child acquires language. And there is no reason not to suppose that the same holds true for the fundamental categories of cognition in man, and perhaps also for a good many other elements of human behavior, less basic but of great consequence in the shaping of the individual and society. Such problems are approachable through experiment. Ethologists conduct such experiments every day. Being cruel experiments, their practice upon human beings (in fact, upon young human beings) is unthinkable. Respect for himself thus compels man to forgo the exploration of some of the very roots of his own nature.

The lengthy controversy over the Cartesian innateness of "ideas," denied by the empiricists, is in a way similar to the more recent one which has divided biologists with regard to the distinction between phenotype and genotype. For the geneticists who introduced it the distinction was fundamental, indispensable to the very definition of the hereditary patrimony; for many biologists not working in genetics it was, on the contrary, a very suspect distinction, in their eyes a device intended to save the postulate of the invariance of the gene. Here, once more, is a recurrence of the conflict between those for whom truth resides only in the concrete object, actually and fully present, and those who look beyond the object for the ideal form it masks. There are but two kinds of scholars, Alain has said: those who love ideas and those who loathe them. In the world of science these two attitudes continue to oppose each other; but both, by their confrontation, are necessary to scientific progress. One can only regret (on their behalf) that this progress, to which those who scorn ideas themselves contribute, invariably decides against them.

In one very important sense, though, the great eighteenth-century empiricists were not wrong. It is perfectly true that in living beings everything, including genetic innateness, comes from experience, whether it be the stereotyped behavior of bees or the innate framework of human cognition. Everything comes from experience; yet not from ongoing current experience, reiterated by each individual with each new generation, but instead, from the experience accumulated by the entire ancestry of the species over the course of its evolution. Only this experience wrung from chance – only those countless trials chastened by selection – could, as with any other organ, have made the central nervous system into an organ adapted to its particular function. As regards the brain: to give a representation of the material world adequate for the performances of the species; to furnish a framework permitting efficient classification of the otherwise unusable data of objective experience; and even, in man, to simulate experience subjectively so as to anticipate its results and prepare action.

It is the powerful development and intensive use of the simulative function that, in my view, characterizes the unique properties of man's brain. And this at the most basic level of the cognitive functions, those on which language rests and

the function of simulation

which it probably reveals only incompletely. Simulation is not an exclusively human function, however. The puppy that manifests its joy at seeing its master getting ready for the daily walk obviously imagines – that is, simulates through anticipation – the discoveries it is about to make, the adventures awaiting it, the exciting perils it will face, but without danger thanks to the reassuring presence of its protector. Later on it will simulate the whole thing again, pell-mell, in a dog's dream.

In animals, as in young children too, subjective simulation appears to be only partially dissociated from neuromotor activity. Play is its outward expression. But in man subjective simulation becomes the superior function par excellence, the creative function. This is what is reflected by the symbolism of language which, transposing and summarizing its operations, recasts it in the form of speech. Whence the fact, underlined by Chomsky, that even in its humblest employments language is almost always innovative: for it translates a subjective experience, a particular simulation that is always new. And in this also human language is totally unlike animal communication. The latter amounts simply to calls and warnings corresponding to a certain number of stereotyped concrete situations. While doubtless capable of fairly precise subjective simulation, the most intelligent animal has no means "to unburden its mind" except by roughly indicating in what *direction* its imagination is turned. Man, on the other hand, is able to give voice to his subjective experiences: the fresh discovery, the creative encounter need not be buried along with him in whom it has been simulated for the first time.

I am sure every scientist must have noticed how his

mental reflection, at the deeper level, is not verbal: to be absorbed in thought is to be embarked upon an *imagined experience*, an experience simulated with the aid of forms, of forces, of interactions which together only barely compose an "image" in the visual sense of the term. Let the attention so concentrate upon the imagined experience as to be oblivious to all else, and I know—for it has happened to me—that one may suddenly find oneself identifying with the object itself, with, say, a molecule of protein. However, it is not then that the significance of the simulated experience comes clear, but only when it has been enunciated symbolically. Indeed, the nonvisual images with which simulation works would be more rightly regarded not as symbols but, if I may so phrase it, as the subjective and abstract "reality" offered directly to imaginary experience.

At any rate, in everyday practice the process of simulation is entirely masked by the utterance which follows it almost immediately and which seems inseparable from thought. But, as we know, numerous observations prove that in man the cognitive functions, even the most complex ones, are not immediately tied in with speech (nor with any other means of symbolic expression). The studies made upon various types of aphasia may be cited in particular. Perhaps the most impressive experiments are the recent ones Sperry conducted with subjects whose two cerebral hemispheres had been separated by surgical cutting of the cross-connecting *corpus callosum*.*
In these subjects the right eye and right hand communi-

---

* J. Levi-Agresti and R. W. Sperry, *Proceedings of the National Academy of Sciences, 61* (1968), 1151.

cate information to and receive it from the left hemisphere only. Thus, an object perceived by the left eye or felt with the left hand is recognized without the subject being able to identify it by name. Now in certain difficult tests that involve matching the three-dimensional shape of an object held in either hand to the flattened-out, two-dimensional picture of that object projected onto a screen, the aphasic right hemisphere proved itself far superior to the "dominant" left hemisphere – not just more accurate, but able to discriminate more rapidly. It is tempting to speculate upon the possibility that the right hemisphere is responsible for an important part, perhaps the more "profound" part, of subjective simulation.

If we are correct in our surmise that thought reposes upon an underlying process of subjective simulation, we must then assume that the high development of this faculty in man is the outcome of an evolution during which natural selection tested the efficacy of the process, its survival value. The very practical terms of this testing have been the success of the concrete action counseled and prepared for by imaginary experimentation. Hence it was on account of its capacity for adequate representation and for accurate foresight *confirmed by concrete experience* that the power of simulation lodged in our early ancestors' central nervous system was propelled to the level reached with Homo sapiens. The subjective simulator could afford to make no mistakes when organizing a panther hunt with the weapons available to Australanthropus, Pithecanthropus or even Homo sapiens of Cro-Magnon times. That is why the innate logical instrument we have inherited from our forebears is so reliable and enables us to "comprehend" events in the world around us, that is, to describe them in symbolic language and to foresee their course, provided the simulator is fed with the necessary elements of information.

As the instrument of intuitive preconception continually enriched by lessons learned from its own subjective experiments, the simulator is the instrument of discovery and of creation. Analysis via language of the logic of its subjective functioning has made possible the formulation of laws of objective logic and the creation of new symbolic instruments such as mathematics. Great thinkers, Einstein among them, have often and justly wondered at the fact that the mathematical entities created by man can so faithfully represent nature even though they owe nothing to experience. Nothing, it is true, to individual and concrete experience; but everything to the virtues of the simulator forged by the vast and bitter experience of our humble ancestors. In systematically setting logic face to face with experience, according to the scientific method, what we are in fact doing is confronting all the experience of our ancestors with that actually facing us.

While we are able to divine the existence of this marvelous instrument, while we know how to translate, through language, the result of its operation, we have no idea of how it works, of its structure. Physiological experimentation has in this regard been mainly unavailing so far. Introspection, despite all its dangers, does tell us somewhat more. There is also the analysis of language, but the process of simulation displayed in language has undergone unknown transformations, and language probably does not exhibit all its operations.

158

There lies the frontier, still almost as impassable for us as it was for Descartes. Not until that barrier has been passed will dualism cease to be a force, and to that extent

the dualist illusion and the presence of the spirit

a truth, in the lives of all of us. We today are no less in the habit of differentiating between brain and mind than they were in the eighteenth century. Objective analysis obliges us to see that this seeming duality within us is an illusion. But it is so well within, so intimately rooted in our being, that nothing could be vainer than to hope to dissipate it in the immediate awareness of subjectivity, or to learn to live emotionally or morally without it. And, besides, why should one have to? What doubt can there be of the presence of the spirit within us? To give up the illusion that sees in it an immaterial "substance" is not to deny the existence of the soul, but on the contrary to begin to recognize the complexity, the richness, the unfathomable profundity of the genetic and cultural heritage and of the personal experience, conscious or otherwise, which together constitute this being of ours: the unique and irrefutable witness to itself.

IX

The Kingdom and the Darkness

WE HAVE already spoken of the day when, venturing beyond the communication of concrete and actual experience, Australanthropus or one of his kin managed to express the content of a subjective experience, of a personal "simulation." On that day a new world was born, the world of ideas; and a new evolution, that of culture, became possible. From there on and for a long time, man's physical evolution must have been intimately connected with and profoundly influenced by the development of the linguistic capacity, which so thoroughly changed the conditions of selection.

the pressures of selection in the evolution of man

Modern man is the product of that evolutionary symbiosis. Viewed otherwise, he is incomprehensible, indecipherable. Every living being is *also* a fossil. Within it, all the way down to the microscopic structure of its proteins, it bears the traces if not the stigmata of its ancestry. This is yet truer of man than of any other animal species by dint of the dual evolution—physical and ideational—that he is heir to.

One may suppose that for hundreds of thousands of

years ideational evolution kept only a short pace ahead of physical evolution, its progress hampered by the meager development of a cortex capable only of anticipating events directly related to immediate survival. Whence the intense selective pressure which was to spur the development of the power of simulation and of the language that conveys its operations. Whence also the astonishing swiftness of this evolution, attested to by fossil skulls.

But as this joint evolution went forward, its ideational component could only tend to greater independence of the restraints which the central nervous system's own development gradually abolished. Owing to this evolution man extended his dominion over the subhuman sphere and suffered less from the dangers it harbored for him. The selective pressure which had guided the first phase of the evolution could then ease, in any case taking on a different character. Now dominating his environment, man had no serious adversary to face other than his own kind. Direct intraspecific strife—mortal strife within his own species—henceforth became one of the principal factors of selection in the human species. In the evolution of animals the phenomenon is extremely rare: we observe today no such thing as intraspecific warfare between distinct races or groups within any given animal species. Among the larger mammals even single combat, frequent between males, seldom leads to the death of the loser. Specialists all agree in thinking that direct strife, Spencer's "struggle for life," has played only a minor role in the evolution of species. This is not so as regards mankind. Somewhere in the human species' development and expansion the point was reached where tribal or racial warfare came to be an important evolutionary factor. It is quite possible that the sudden disappearance of Neanderthal man was the work of our ancestor Homo sapiens. It was not to be the last performance of its kind: genocides abound in recorded history.

In what direction did this selective pressure push human evolution? Of course it favored the expansion of races more generously endowed than others with intelligence, imagination, will, and ambition. But it must also have favored cohesion within the horde and group aggressiveness more than lone courage, respect for the tribal law more than individual initiative.

This is a simplified outline indeed, and I am quite willing to hear it criticized. I do not mean to divide human evolution into two distinct phases. I have only tried to enumerate the main selective pressures that must have counted heavily in man's cultural and also physical evolution. The important point is that during those hundreds of thousands of years, cultural evolution could not help but affect physical evolution; in man more than in any other animal—and owing precisely to its infinitely greater autonomy—it is *behavior* that *orients* selective pressure. And once that behavior ceased to be primarily automatic and became cultural, cultural traits themselves inevitably exerted their pressure upon the evolution of the genome.

This, up until the moment when the accelerating pace of cultural evolution was to split completely away from that of the genome.

Within the framework of modern societies this split is obviously total. In them selection has been done away with. Or at least, there is no longer anything "natural" about it in the Darwinian sense. To the extent that selection is still operative in our midst, it does not favor the

dangers of genetic degradation in modern societies

"survival of the fittest" – that is to say, in more modern terms, the *genetic* survival of the "fittest" through a more numerous progeny. Intelligence, ambition, courage, and imagination are still factors of success in modern societies, to be sure; but of *personal*, not *genetic* success, the only kind that matters for evolution. No, the situation is the reverse: statistics, as everybody knows, show a negative correlation between the intelligence quotient (or cultural level) and the average number of children per couple. These same statistics demonstrate, meanwhile, that there is a high positive correlation of intelligence quotients between marital partners. A dangerous situation, this, which could gradually edge the highest genetic potential toward concentration within an élite, a shrinking élite in relative numbers.

This is not all. Until not so very long ago, even in relatively "advanced" societies, the weeding out of the physically and also mentally least fit was automatic and ruthless. Most of them did not reach the age of puberty. Today many of these genetic cripples live long enough to reproduce. Thanks to the progress of scientific knowledge and the social ethic, the mechanism which used to protect the species from degeneration (the inevitable result when natural selection is suspended) now functions hardly at all, save where the defect is uncommonly grave.

For coping with these dangers, often signaled to our attention, there are occasional promises of remedies expected from the current advances in molecular genetics. This illusion, spread about by a few superficial minds, had better be disposed of. No doubt it will be possible to palliate certain genetic flaws, *but only in the afflicted individual*, not in his posterity. Not only does modern molecular genetics give us *no means whatsoever* for acting upon the ancestral heritage in order to improve it with new features – to create a genetic "superman" – but it reveals the vanity of any such hope: the genome's microscopic proportions today and probably forever rule out manipulation of this sort. Science fiction's chimerical schemes set aside, the only means for "improving" the human species would be to introduce a deliberate and severe selection. Who will want – who will dare to employ it?

Conditions of nonselection (or of selection-in-reverse) like those reigning in the advanced societies are a definite peril to the species. For it to become very serious, however, would take quite a while: say ten or fifteen generations, or several centuries. And there are far more grave and more urgent dangers threatening modern societies already.

Here, I am not referring to the population explosion, to the destruction of the natural environment, nor even to the stock pile of megatons of nuclear power; but to a more insidious and much more deep-seated evil: one that besets the spirit. One that was begot of the sharpest turning point ever taken in the evolution of ideas. An evolution, moreover, which continues and accelerates constantly in the same direction, ever increasing that bitter distress of the soul.

The impact of his prodigious attainments in all areas of knowledge over the past three centuries is forcing man to make a heart-rending revision in his concept of himself and his relation to the world, a concept which had become rooted in him through tens of thousands of years.

The whole of it, however—the spirit's disorder like our nuclear might—is the outcome of one simple idea: that nature is objective, that the systematic confronting of logic and experience is the sole source of true knowledge. It is hard to understand how, in the kingdom of ideas, this one, so simple and so clear, failed to come fully through until a hundred thousand years after the emergence of Homo sapiens; why it never cropped up in some of the loftiest civilizations, such as the Chinese, which had to learn it from the West; or why, in the West itself, nearly twenty-five hundred years were needed, from Thales and Pythagoras to Galileo, Bacon, and Descartes, before it shook loose from its encapsulation within the practice of the mechanical arts.

the selection of ideas	For a biologist it is tempting to draw a parallel between the evolution of ideas and that of the biosphere. For while the ab-

stract kingdom stands at a yet greater distance above the biosphere than the latter does above the nonliving universe, ideas have retained some of the properties of organisms. Like them, they tend to perpetuate their structure and to breed; they too can fuse, recombine, segregate their content; indeed they too can evolve, and in this evolution selection must surely play an important role. I shall not hazard a theory of the selection of ideas. But one may at least try to define some of the principal factors involved in it. This selection must necessarily operate at two levels: that of the mind itself and that of performance.

The performance value of an idea depends upon the change it brings to the behavior of the person or the group that adopts it. The human group upon which a given idea confers greater cohesiveness, greater ambition, and greater self-confidence thereby receives from it an added power to expand which will insure the promotion of the idea itself. Its capacity to "take," the extent to which it can be "put over" has little to do with the amount of objective truth the idea may contain. The important thing about the stout armature a religious ideology constitutes for a society is not what goes into its structure, but the fact that this structure is accepted, that it gains sway. So one cannot well separate such an idea's power to spread from its power to perform.

The "spreading power" – the infectivity, as it were – of ideas, is much more difficult to analyze. Let us say that it depends upon preexisting structures in the mind, among them ideas already implanted by culture, but also undoubtedly upon certain innate structures which we are hard put to identify. What is very plain, however, is that the ideas having the highest invading potential are those that *explain* man by assigning him his place in an immanent destiny, in whose bosom his anxiety dissolves.

During entire aeons a man's lot was identical with that of the group, of the tribe he belonged to and outside of which he could not survive. The tribe, for its part, was able to survive and defend itself only through its cohesion. Whence the extreme force of inward coercion exerted by the laws that organized and guaranteed this cohesion. A man might perhaps infringe them; it is not likely that any man ever dreamed of denying them. Given the

the need for an explanation

immense selective importance such social structures perforce assumed over such vast stretches of time, it is difficult not to believe that they must have made themselves felt upon the genetic evolution of the innate categories of the human brain. This evolution must not only have facilitated acceptance of the tribal law, but created the *need* for the mythical explanation which gave it foundation and sovereignty. We are the descendants of such men. From them we have probably inherited our need for an explanation, the profound disquiet which goads us to search out the meaning of existence. That same disquiet has created all the myths, all the religions, all the philosophies, and science itself.

That this imperious need develops spontaneously, that it is inborn, inscribed somewhere in the genetic code, strikes me as beyond doubt. Outside the human species, nowhere in the animal kingdom does one find such highly differentiated social organizations save it be among certain insects: ants, termites, bees. But the stability of the social insects' institutions owes next to nothing to cultural heritage, virtually everything to genetic transmission. Social behavior, with them, is entirely innate, automatic.

Man's social institutions, purely cultural, cannot ever attain a like stability; besides, who would wish for such a thing? The invention of myths and religions, the construction of vast philosophical systems—they are the price this social animal has had to pay in order to survive without having to yield to pure automatism. But to anchor the social structure, the cultural tradition, all by itself, would not have been reliable enough, strong enough. That heritage needed a genetic support to make it into something the mind could not do without. How else account for the fact that in our species the religious phenomenon is invariably at the base of social structure? How else explain that, throughout the immense variety of our myths, our religions, and philosophical ideologies, the same essential "form" always recurs?

It is readily seen that the "explanations" meant to give foundation to the law while assuaging man's anxiety

mythic and metaphysical ontogenies

are all narrations of past events, "stories"–or "histories"–that are (in the philosophical sense) really "ontogenies." Primitive myths almost all tell of more or less divine heroes whose deeds explain the origins of the group and base its social structure upon immutable traditions; one does not remake history. The great religions are of similar form, resting on the story of the life of an inspired prophet who, if not himself the founder of all things, represents that founder, speaks for him, and recounts the history of mankind as well as its destiny. Of all the great religions Judeo-Christianity is probably the most "primitive," since its strictly historicist structure is directly plotted upon the saga of a Bedouin tribe before being enriched by a divine prophet. Buddhism, on the contrary, more highly differentiated, has recourse in its original form to Karma alone, the transcending law governing individual destiny. Buddhism is more a story of souls than of men.

From Plato to Hegel and Marx, the great philosophical systems all propose at once explanatory and normative

168

ontogenies. It is true that with Plato the course is down-hill rather than ascending. He sees in history only the gradual corruption of ideal forms, and in the *Republic* the aim is to reinstate the past, to move backward in time.

For Marx as for Hegel history unfolds according to an immanent, necessary, and favorable plan. The immense influence of Marxist ideology does not derive alone from its promise of liberation for man, but also, and probably mainly, from its ontogenic structure, the explanation it provides, sweeping and in detail, of past, present, and future history. However, limited to human history, even though decked with the certainties of "science," historical materialism was yet incomplete. It needed the addition of dialectical materialism in order to become the total inter-pretation the mind needs: human history connects with that of the cosmos to obey the same eternal laws.

If it is true that the need for a complete explanation is innate, that its absence begets a profound ache within; if the only form of explanation capable of putting the soul at ease is that of a total history which discloses the meaning of man by assigning him a neces-sary place in nature's scheme; if, to appear genuine, meaningful, soothing, the "explanation" must blend into the long animist tradi-

the breakdown of the old covenant and the modern soul's distress

tion, then we understand why it took so many thousands of years for the kingdom of ideas to be invaded by the one according to which objective knowledge is the *only* authentic source of truth.

Cold and austere, proposing no explanation but imposing an ascetic renunciation of all other spiritual fare, this idea was not of a kind to allay anxiety, but aggravated it instead. By a single stroke it claimed to sweep away the tradition of a hundred thousand years, which had become one with human nature itself. It wrote an end to the ancient animist covenant between man and nature, leaving nothing in place of that precious bond but an anxious quest in a frozen universe of solitude. With nothing to recommend it but a certain puritan arrogance, how could such an idea win acceptance? It did not; it still has not. It has however commanded recognition; but that is because, solely because, of its prodigious power of performance.

In the course of three centuries science, founded upon the postulate of objectivity, has conquered its place in society — in men's practice, but not in their hearts. Modern societies are built upon science. They owe it their wealth, their power, and the certitude that tomorrow far greater wealth and power still will be ours if we so wish. But there is this too: just as an initial "choice" in the biological evolution of a species can be binding upon its entire future, so the choice of scientific *practice,* an unconscious choice in the beginning, has launched the evolution of culture on a one-way path; onto a track which nineteenth-century scientism saw leading infallibly upward to an empyrean noon hour for mankind, whereas what we see opening before us today is an abyss of darkness.

Modern societies accepted the treasures and the power that science laid in their laps. But they have not accepted — they have scarcely even heard — its profounder message: the defining of a new and unique source of truth, and the demand for a thorough revision of ethical premises, for a

total break with the animist tradition, the definitive abandonment of the "old covenant," the necessity of forging a
new one. Armed with all the powers, enjoying all the rich-
es they owe to science, our societies are still trying to
live by and to teach systems of values already blasted at
the root by science itself.

No society before ours was ever rent by contradictions
so agonizing. In both primitive and classical cultures the
animist tradition saw knowledge and values stemming
from the same source. For the first time in history a
civilization is trying to shape itself while clinging desperately to the animist tradition to justify its values, and at
the same time abandoning it as the source of knowledge,
of *truth*. For their moral bases the "liberal" societies of
the West still teach – or pay lip-service to – a disgusting
farrago of Judeo-Christian religiosity, scientistic progressism, belief in the "natural" rights of man, and utilitari-
an pragmatism. The Marxist societies still profess the
materialist and dialectical religion of history; on the face
of it a more solid moral framework than the liberal societies boast, but perhaps more vulnerable by virtue of the
very rigidity that has made its strength up until now.
However this may be, all these systems rooted in animism exist at odds with objective knowledge, face away
from truth, and are strangers and fundamentally *hostile* to
science, which they are pleased to make use of but for
which they do not otherwise care. The divorce is so great,
the lie so flagrant, that it afflicts and rends the conscience
of anyone provided with some element of culture, a little
intelligence, and spurred by that moral questioning which
is the source of all creativity. It is an affliction, that is to
say, for all those among mankind who bear or will come to

171

bear the responsibility for the way in which society and culture shall evolve.

What ails the modern spirit is this lie gripping man's moral and social nature at the very core. It is this ailment, more or less confusedly diagnosed, that provokes the fear if not the hatred—in any case the estrangement—felt toward scientific culture by so many people today. Their aversion, when openly expressed, usually directs itself at the technological by-products of science: the bomb, the destruction of nature, the soaring population. The easy reply, of course, is that technology and science are not the same thing, and moreover that the use of atomic energy will soon be vital to mankind's survival; that the destruction of nature denotes a faulty technology rather than too much of it; and that the population soars because children by the millions are saved from death every year. Are we to go back to letting them die?

Confusing the symptoms of the disorder with its underlying cause, this is a superficial reply. Indeed, it merely begs the question. For behind the protest is the denial of the essential message of science. The fear is the fear of sacrilege: of outrage to values. A wholly justified fear. It is perfectly true that science outrages values. Not directly, since science is no judge of them and *must* ignore them; but it subverts every one of the mythical or philosophical ontogenies upon which the animist tradition, from the Australian aborigines to the dialectical materialists, has made all ethics rest: values, duties, rights, prohibitions.

If he accepts this message—accepts all it contains— then man must at last wake out of his millenary dream; and in doing so, wake to his total solitude, his fundamental isolation. Now does he at last realize that, like a gypsy, he lives on the boundary of an alien world. A world that is deaf to his music, just as indifferent to his hopes as it is to his suffering or his crimes.

But henceforth who is to define crime? Who shall decide what is good and what is evil? All the traditional systems have placed ethics and values beyond man's reach. Values did not belong to him; he belonged to them. He now knows that they are his and his alone, and they no sooner come into his possession than lo! they seem to melt into the world's uncaring emptiness. It is then that modern man turns toward science, or rather against it, finally measuring its terrible capacity to destroy not only bodies but the soul itself.

Where is the remedy? Must one adopt the position once and for all that objective truth and the theory of values constitute eternally separate, mutually impenetrable domains? This is the attitude taken by a great number of modern thinkers, whether writers, or philosophers, or indeed scientists. For the vast majority of men, whose anxiety it can only perpetuate and worsen, this attitude I believe will not do; I also believe it is absolutely mistaken, and for two essential reasons.

values and knowledge

First, and obviously, because values and knowledge are always and necessarily associated in action just as in discourse.

Second, and above all, because *the very definition of "true" knowledge reposes in the final analysis upon an ethical postulate.*

Each of these two points demands some brief clarification.

Ethics and knowledge are inevitably linked in and through action. Action brings knowledge and values *simultaneously* into play, or into question. All action signifies an ethic, serves or disserves certain values; or constitutes a choice of values, or pretends to. On the other hand, knowledge is necessarily implied in all action, while reciprocally, action is one of the two necessary sources of knowledge.

In an animist system the interpenetration of ethics and knowledge creates no conflict, since animism avoids any basic distinction between these two categories: it sees them as two aspects of the same reality. The idea of a social ethic founded upon the purportedly "natural" rights of man reflects this outlook, also displayed, but much more systematically and emphatically, in the attempts to delineate the ethics implicit in Marxism.

The moment one makes objectivity the *conditio sine qua non* of true knowledge, a radical distinction, indispensable to the very search for truth, is established between the domains of ethics and of knowledge. Knowledge in itself is exclusive of all value judgment (all save that of "epistemological value") whereas ethics, in essence *nonobjective*, is forever barred from the sphere of knowledge.

It is in effect this radical distinction, laid down as an axiom, that created science. I am inclined to add that if this unprecedented event in the history of culture took place in the Christian West rather than in some other civilization, it was perhaps thanks, in part, to the fundamental distinction drawn by the Church between the domains of the sacred and the profane. Not only did this distinction allow science to pursue its own way (provided it did not trespass upon the realm of the sacred); it prepared the mind for the much more radical distinction posed by the principle of objectivity. Westerners often have trouble understanding that for certain religions there neither is nor can be any distinguishing between sacred and profane: for Hinduism, everything comes within the bounds of the sacred; the very concept of "profane" is incomprehensible.

Let us, however, return to our main point. The postulate of objectivity, denouncing the "old covenant," at the same stroke prohibits any confusion of value judgments with judgments arrived at through knowledge. Yet the fact remains that these two categories inevitably unite in the form of action, discourse included. In order to abide by our principle we shall therefore take the position that no discourse or action is to be considered meaningful, *authentic*, unless – or only insofar as – it makes explicit and preserves the distinction between the two categories it combines. Thus defined, the concept of authenticity becomes the common ground where ethics and knowledge meet again; where values and truth, associated but not interchangeable, reveal their full significance to the attentive man alive to their resonance. In return, *inauthentic* discourse, where the two categories are jumbled, can lead only to the most pernicious nonsense, to perhaps unwitting but nonetheless criminal lies.

(It is in "political" discourse, clearly, that this hazardous amalgamation is most consistently and systematically practiced. And not by professional politicians alone. Scientists themselves, outside their field, often prove dangerously incapable of distinguishing between the categories of values and of knowledge.)

Animism, we said earlier, neither wants nor for that matter is able to set up an absolute discrimination between value judgments and statements based upon knowledge; for having once assumed that there is an intention, however carefully disguised, present in the universe, what would be the sense of such a distinction? In an objective system the very opposite holds: any mingling of knowledge with values is unlawful, *forbidden.* But – and here is the crucial point, the logical link which at their core weds knowledge and values together – this prohibition, this "first commandment" which ensures the foundation of objective knowledge, is not itself objective. It cannot be objective: it is an ethical guideline, a rule for conduct. True knowledge is ignorant of values, but it cannot be grounded elsewhere than upon a value judgment, or rather upon an *axiomatic* value. It is obvious that the positing of the principle of objectivity as the condition of true knowledge *constitutes an ethical choice and not a judgment arrived at from knowledge, since, according to the postulate's own terms, there cannot have been any "true" knowledge prior to this arbitral choice.* In order to establish the *norm* for knowledge the objectivity principle defines a *value:* that value is objective knowledge itself. Thus, assenting to the principle of objectivity one announces one's adherence to the basic statement of an ethical system, one asserts *the ethic of knowledge.*

Hence it is from the ethical choice of a primary value that knowledge starts. The ethic of knowledge thereby differs radically from animist ethics, which all claim to be based

**the ethic
of knowledge**

upon the "knowledge" of immanent laws, religious or "natural," which are supposed to assert themselves over man. The ethic of knowledge does not obtrude itself upon man; *on the contrary, it is he who prescribes it to himself*, making of it the *axiomatic* condition of authenticity for all discourse and all action. The *Discours de la Méthode* proposes a normative epistemology, but it must also be read above all as a moral meditation, as a spiritual exercise.

Authentic discourse in its turn lays the foundation of science, and returns to the hands of man the immense powers that enrich and imperil him today. Modern societies, woven together by science, living from its products, have become as dependent upon it as an addict on his drug. They owe their material wherewithal to this fundamental ethic upon which knowledge is based, and their moral weakness to those value-systems, devastated by knowledge itself, to which they still try to refer. The contradiction is deadly. It is what is digging the pit we see opening under our feet. The ethic of knowledge that created the modern world is the only ethic compatible with it, the only one capable, once understood and accepted, of guiding its evolution.

Understood and accepted – could it be? If it is true, as I believe, that the fear of solitude and the need for a complete and binding explanation are inborn – that this heritage from the very remote past is not only cultural but probably genetic too – can one imagine such an ethic as this, austere, abstract, proud, calming that fear, satisfying that need? I do not know. But it may not be altogether impossible. Perhaps, even more than an explanation which the ethic of knowledge cannot supply, it is to rise above himself that man craves. The abiding power of the great socialist dream, still present in men's hearts, would indeed seem to suggest it. No system of values can be said to constitute a true ethic unless it proposes an ideal reaching beyond the individual and transcending the self to the point even of justifying self-sacrifice, if need be.

By the very loftiness of its ambition the ethic of knowledge might perhaps satisfy this urge in man to project toward something higher. It sets forth a transcendant value, true knowledge, and invites him not to use it self-servingly but henceforth to enter into its service from deliberate and conscious choice. At the same time it is also a humanism, for in man it respects the creator and repository of that transcendence.

The ethic of knowledge is also in a sense "knowledge of ethics," a clear-sighted appreciation of the urges and passions, the requirements and limitations of the biological being. It is able to confront the animal in man, to view him not as absurd but strange, precious in his very strangeness: the creature who, belonging simultaneously to the animal kingdom and the kingdom of ideas, is simultaneously torn and enriched by this agonizing duality, alike expressed in art and poetry and in human love.

Conversely, the animist systems have to one degree or another preferred to ignore, to denigrate or bully biological man, and to instill in him an abhorrence or terror of certain traits inherent in his animal nature. The ethic of knowledge, on the other hand, encourages him to honor and assume this heritage, knowing the while how to dominate it when necessary. As for the highest human qualities, courage, altruism, generosity, creative ambition, the ethic of knowledge both recognizes their sociobiological origin and affirms their transcendent value in the service of the ideal it defines.

Finally, the ethic of knowledge is, in my view, the one at once rational and resolutely idealistic attitude upon which a real socialism might be built. To the young in spirit that great vision of the nineteenth century continues to beckon with grievous intensity. Grievous because of the betrayals this ideal has suffered, and because of the crimes committed in its name. It is tragic, but was perhaps inevitable, that this profound aspiration had to find its philosophical doctrine in the form of an animist ideology. Looking back, how well one sees that, from the time of its birth, historical messianism based on dialectical materialism contained the seeds of all the woe later generations were to harvest. Perhaps more than the other animisms, historical materialism rests upon a total confusion of the categories of value and knowledge. This very confusion permits it, in a travesty of authentic discourse, to proclaim that it has "scientifically" established the laws of history, which man has no choice or duty but to obey if he does not wish to sink away into nothingness.

the ethic of knowledge and the socialist ideal

This illusion, which is merely puerile when it is not fatal, must be given up once and for all. How can an authentic socialism ever be constructed upon an ideology inauthentic to the marrow, ludicrous parody of that very science upon which it claims to stand—most sincerely, in the minds of its adepts. Socialism's one hope is not in a "revision" of the ideology that has been dominating it for better than a century, but in completely throwing that ideology over.

Where then shall we find the source of truth and the moral inspiration for a really *scientific* socialist humanism, if not in the sources of science itself, in the ethic upon which knowledge is founded, and which by free choice makes knowledge the supreme value – the measure and warrant for all other values? An ethic which bases moral responsibility upon the very freedom of that axiomatic choice. Accepted as the foundation for social and political institutions, hence as the measure of their authenticity, their value, only the ethic of knowledge could lead to socialism. It prescribes institutions dedicated to the defense, the extension, the enrichment of the transcendent kingdom of ideas, of knowledge, and of creation – a kingdom which is within man, where progressively freed both from material constraints and from the deceitful servitudes of animism, he could at last live authentically, protected by institutions which, seeing in him the subject of the kingdom and at the same time its creator, could be designed to serve him in his unique and precious essence.

A utopia. Perhaps. But it is not an incoherent dream. It is an idea that owes its force to its logical coherence alone. It is the conclusion to which the search for authenticity necessarily leads. The ancient covenant is in pieces; man knows at last that he is alone in the universe's unfeeling immensity, out of which he emerged only by chance. His destiny is nowhere spelled out, nor is his duty. The kingdom above or the darkness below: it is for him to choose.

Appendixes

1
Structure of Proteins

PROTEINS ARE macromolecules constituted by the linear polymerization of compounds called amino acids. The general structure of the "polypeptide" chain resulting from this polymerization is shown in the following drawing.

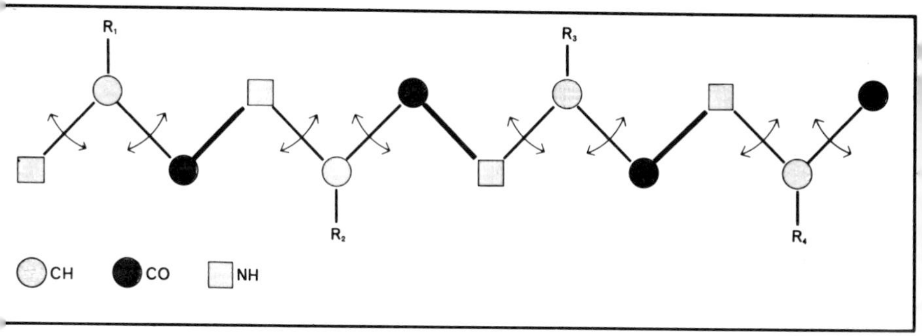

The white and black circles and the white squares correspond to various groupings of atoms ($\bigcirc$ = CH; $\bullet$ = CO; $\square$ = NH) while the letters R_1, R_2, etc. represent different organic residues. The twenty amino acid residues that are the universal constituents of proteins are shown in Table I.

183

TABLE I

AMINO ACID RESIDUES

A. Hydrophobic

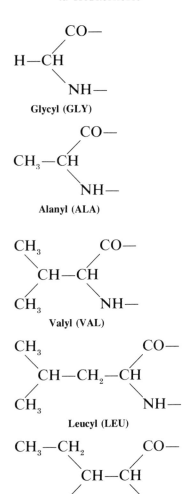

Glycyl (GLY)

Alanyl (ALA)

Valyl (VAL)

Leucyl (LEU)

Isoleucyl (ILEU)

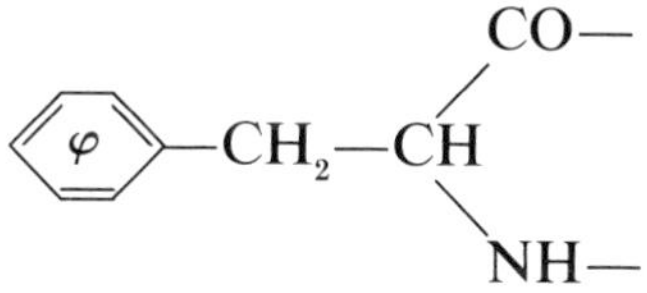

Structure of Proteins

Phenylalanyl (PHE)

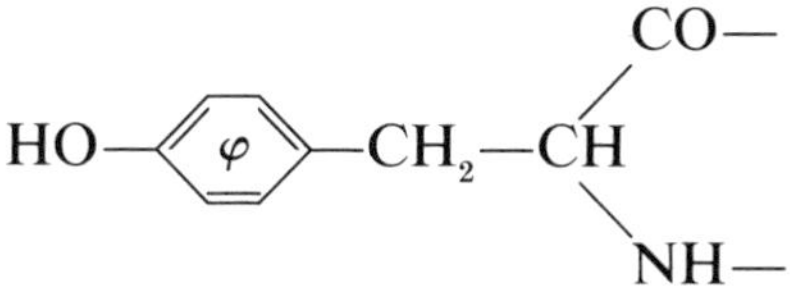

Tyrosyl (TYR)

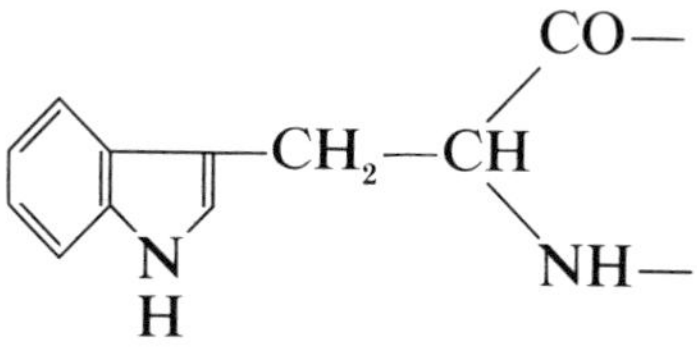

Tryptophanyl (TRY)

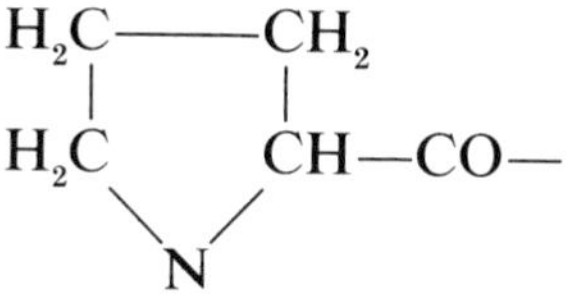

Prolyl (PRO)

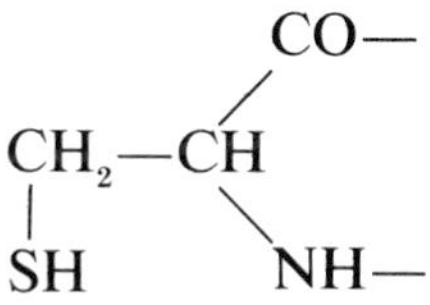

Cysteinyl (CYS)

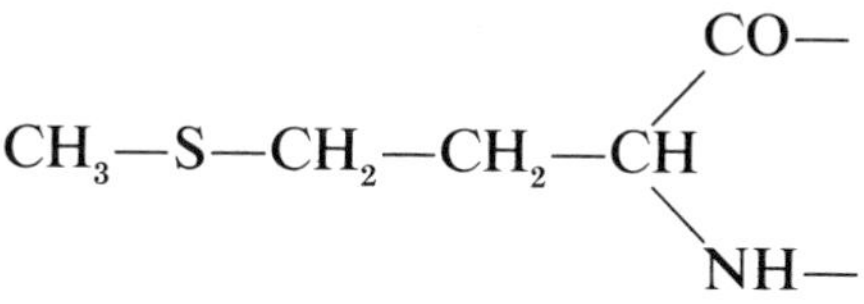

Methionyl (MET)

185

TABLE I

AMINO ACID RESIDUES

B. HYDROPHILIC

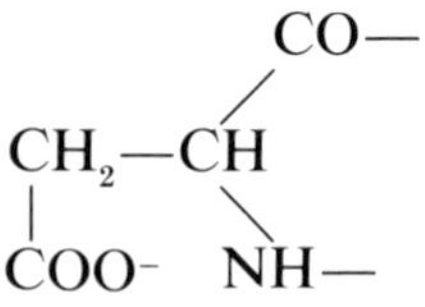

$$CH_2-CH \overset{\displaystyle CO-}{\underset{\displaystyle NH-}{}}$$

$$COO^-$$

Aspartyl (ASP)

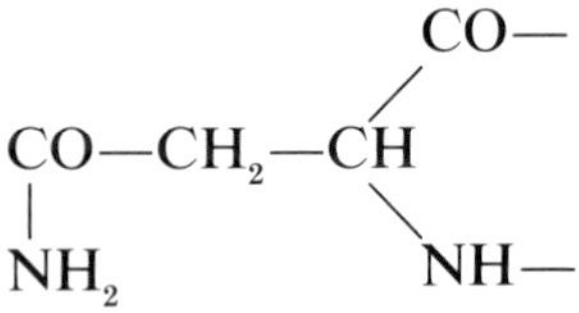

$$CO-CH_2-CH \overset{\displaystyle CO-}{\underset{\displaystyle NH-}{}}$$

$$NH_2$$

Asparagyl (ASPN)

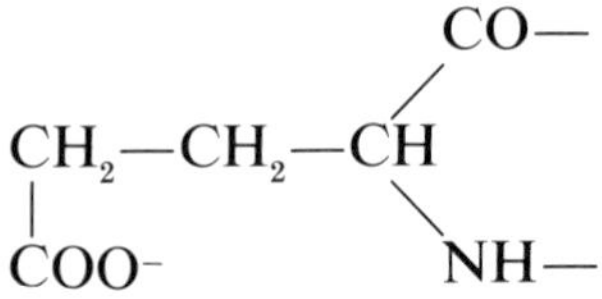

$$CH_2-CH_2-CH \overset{\displaystyle CO-}{\underset{\displaystyle NH-}{}}$$

$$COO^-$$

Glutamyl (GLU)

$$CO-CH_2-CH_2-CH \overset{\displaystyle CO-}{\underset{\displaystyle NH-}{}}$$

$$NH_2$$

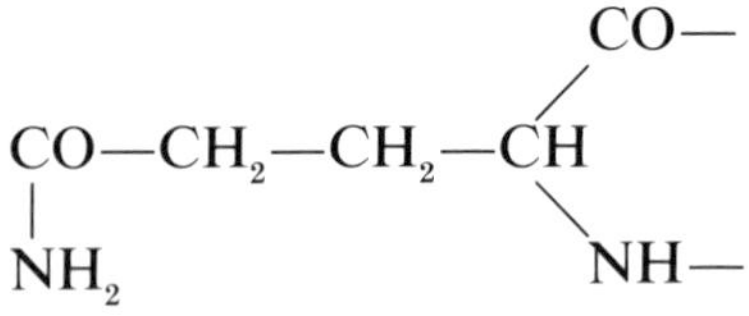

Glutaminyl (GLUN)

The chain includes three kinds of bonds between atoms or groups of atoms.

1. Between white circle and black circle (CH—CO);
2. Between white circle and white square (CH—NH);
3. Between black circle and white square (CO—NH).

186

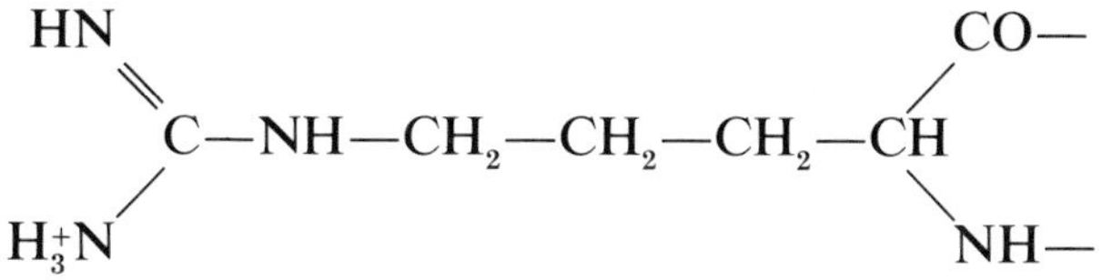

$$\begin{array}{c} HN \\ \parallel \\ C-NH-CH_2-CH_2-CH_2-CH \\ H_3^+N \end{array} \quad \begin{array}{c} CO- \\ \\ NH- \end{array}$$

Arginyl (ARG)

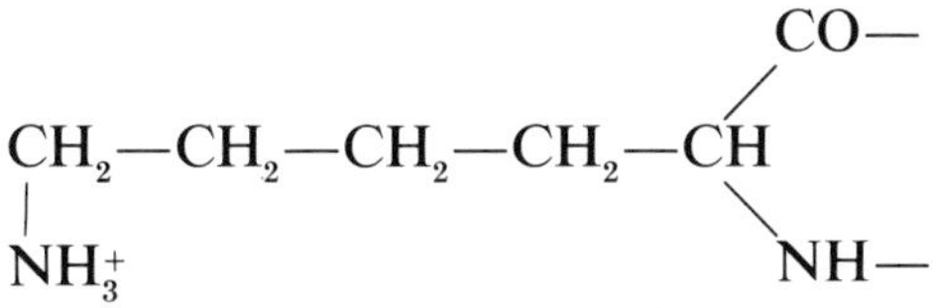

$$\begin{array}{c} CH_2-CH_2-CH_2-CH_2-CH \\ NH_3^+ \end{array} \quad \begin{array}{c} CO- \\ \\ NH- \end{array}$$

Lysyl (LYS)

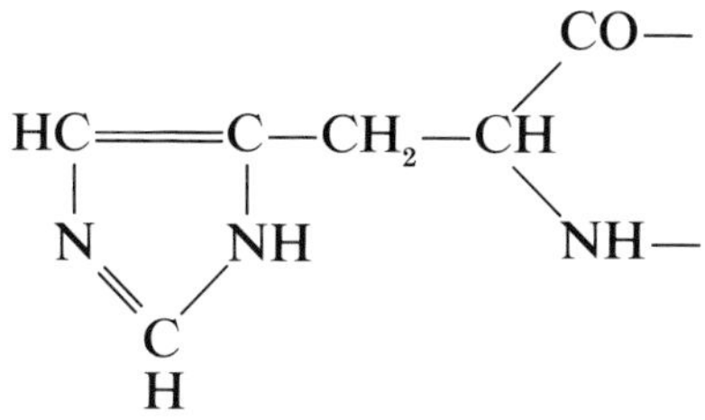

$$\begin{array}{c} HC=\!\!=\!\!C-CH_2-CH \\ N \quad NH \\ \diagdown \; \diagup \\ C \\ H \end{array} \quad \begin{array}{c} CO- \\ \\ NH- \end{array}$$

Histidyl (HIS)

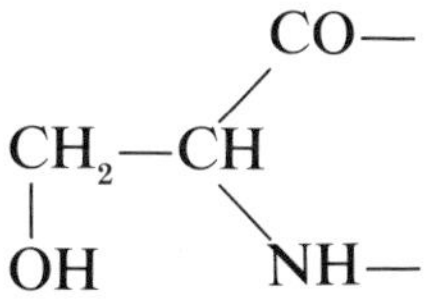

$$\begin{array}{c} CH_2-CH \\ OH \end{array} \quad \begin{array}{c} CO- \\ \\ NH- \end{array}$$

Seryl (SER)

$$\begin{array}{c} CH_3-CH-CH \\ OH \end{array} \quad \begin{array}{c} CO- \\ \\ NH- \end{array}$$

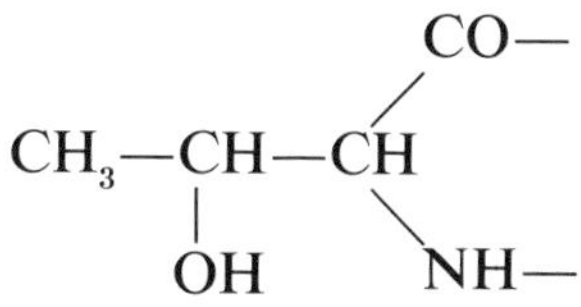

Threonyl (THR)

The last bond, known as the "peptide bond" (represented by heavy lines in the drawing on p. 183) is rigid: the atoms it connects are held immobile with respect to each other. The two other bonds, on the contrary, permit the linked atoms to rotate with respect to each other (dotted arrows in the drawing). This in turn permits the polypeptide fiber to bundle by folding in an extremely varied and flexible way. Only the room taken up by the atoms (notably those that constitute the residues R_1, R_2, etc.) places any limitation upon the fiber's possibilities for folding.

However (see p. 90), in native globular proteins all the molecules of a given chemical species (defined by the *sequence* of the residues in the chain) adopt the same bundled shape. Figure 5, below, diagrammatically illustrates the complex and seemingly incoherent arrangement of the polypeptide chain in the enzyme papain.

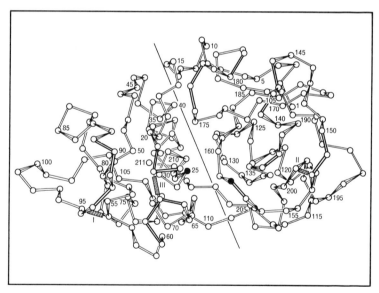

Fig. 5. Schematic representation of the folding of the peptide chain in the papain molecule. J. Drenth, J. N. Jansonius, R. Koekoek, H. M. Swen, and B. G. Wolthers, *Nature, 218* (1968), 929-32.

2
Nucleic
Acids

NUCLEIC ACIDS are macromolecules resulting from the linear polymerization of compounds called "nucleotides." The latter are formed through the association of a sugar with a nitrogen-containing base, on the one hand, and on the other with a phosphoryl group. The polymerization occurs through the intermediary of phosphoric groups which link each sugar residue to the one before and the one after, thus forming a "polynucleotide" chain.

In DNA (deoxyribonucleic acid) are found four nucleotides which differ in the structure of the constituent nitrogen-containing base. These four bases, adenine, guanine, cytosine, and thymine, are usually abbreviated as A, G, C, and T. They are the letters of the genetic alphabet. For steric reasons the adenine (A) in DNA tends to form a spontaneous noncovalent association (see p. 54) with thymine (T), while guanine (G) associates with cytosine (C).

DNA is made up of *two* polynucleotide strands joined by means of these specific noncovalent bonds. In the double strand, A of one strand is linked to T in the other; G to C; T to A; and C to G. The two strands are therefore *complementary*.

This structure is represented diagrammatically in the figure following. In it the pentagons stand for sugar residues, the black circles for the phosphorus atoms which ensure the continuity of both chains; the squares marked A, T, G, and C represent

189

the bases connected into pairs (A – T; G – C; T – A; C – G) by noncovalent interactions, indicated by the dotted lines. The structure can accommodate every possible sequence of pairs. It is not limited as to length.

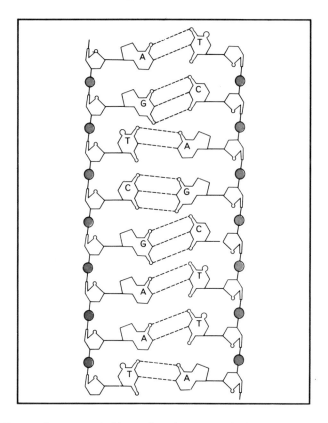

The *replication* of this molecule proceeds by the separation of the duplex, followed by the reconstitution, nucleotide by nucleotide, of the two complements. This is shown — in a simplified manner and confining ourselves to four pairs — in the drawing following.

The two molecules thus synthesized each contain one

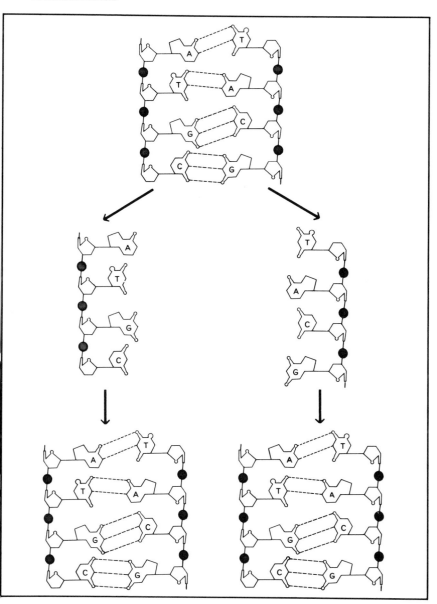

strand of the parent molecule and a strand newly formed by specific nucleotide-by-nucleotide pairing. These two molecules are identical to each other and to the original molecule. Such is the mechanism, very simple in principle, of replicative invariance.

Mutations result from the various kinds of accidents which may alter this microscopic mechanism. The chemistry behind some mutations is today fairly well understood. For example, the substitution of one nucleotide pair for another is accounted for by the fact that the nitrogen-containing bases, besides their "normal" state, can exceptionally and temporarily adopt a form in which their capacity for specific base pairing is so to speak "reversed" (thus, in its "exceptional" state, base C pairs with A rather than with G, and so on). Chemical agents are known which considerably augment the probability, that is to say the frequency, of these "illicit" pairings. These agents are powerful "mutagens."

Other chemical agents, able to wedge themselves *between* the nucleotides in a DNA strand, deform it and thereby induce such accidents as the deletion or addition of one or several extra nucleotides.

Finally, ionizing radiations (X rays and cosmic rays) provoke, *inter alia*, various deletions or "garblings."

3
The Genetic
Code

THE STRUCTURE and properties of a protein are defined by the
sequence (the linear order) of the amino acid residues in the
polypeptide (cf. p. 92). This sequence is itself determined by
that of the nucleotides in a segment of DNA strand. The genetic code (*strictu sensu*) is the rule which prescribes, given
polynucleotide sequence, the corresponding polypeptide sequence.

Since there are twenty amino acids to specify and at the
same time only four "letters" (four nucleotides) in the DNA
alphabet, several nucleotides are required for the specifying of
each amino acid. The code in fact reads in "triplets": each
amino acid is specified by a sequence of *three nucleotides*. The
general features of the code are summarized in Table II, on p.
194.

It is to be noted at once that the translation machinery does
not make direct use of the DNA nucleotide sequences themselves but of a working copy formed by the "transcription" of
one of the two strands into a one-stranded polynucleotide
called "messenger ribonucleic acid" (messenger RNA). The
RNA polynucleotides differ from the DNA nucleotides in a
few details of structure, notably the substitution of the base
uracil (U) for the base thymine (T). Since messenger RNA
serves directly as template for the sequential assembly of the
amino acids which are to make up the polypeptide, the code,

TABLE II

THE GENETIC CODE

I	II	U	C	A	G	III
		PHE	SER	TYR	CYS	U
		PHE	SER	TYR	CYS	C
	U	LEU	SER	nonsense	nonsense	A
		LEU	SER	nonsense	TRY	G
		LEU	PRO	HIS	ARG	U
		LEU	PRO	HIS	ARG	C
	C	LEU	PRO	GLUN	ARG	A
		LEU	PRO	GLUN	ARG	G
		ILEU	THR	ASPN	SER	U
		ILEU	THR	ASPN	SER	C
	A	ILEU	THR	LYS	ARG	A
		MET	THR	LYS	ARG	G
		VAL	ALA	ASP	GLY	U
		VAL	ALA	ASP	GLY	C
	G	VAL	ALA	GLU	GLY	A
		VAL	ALA	GLU	GLY	G

In this table the first letter for each triplet is read in the vertical column on the left; the second letter in the horizontal row at the top; the third in the vertical column on the right. The names of the amino acid residues are given in the abbreviations indicated in Table I, pages 184−7.

shown in Table II, is here written out in the RNA rather than the DNA alphabet.

We see that for most of the amino acids there exist several different notations in the form of nucleotide "triplets." With a

The Genetic Code

four-letter alphabet $4^3 = 64$ three-letter "words" can be formed; there are however only 20 residues to be specified.

On the other hand three triplets (UAA, UAG, UGA) are labeled "Nonsense" because they do not designate any amino acid. They do nevertheless play an important role as punctuation signals (at the beginning or end) in reading the nucleotide sequence.

The actual mechanism of translation is complex; numerous macromolecular constituents are involved in it. A familiarity with this mechanism is not indispensable to an understanding of the text. It will be enough to say a few words about the intermediates that hold the key to the translation process. These intermediates are the so-called "transfer" RNA molecules. These contain:

1. A group which "accepts" amino acids; special enzymes recognize, on the one hand an amino acid, on the other hand a particular transfer RNA, and catalyze the covalent association of the amino acid with the RNA molecule.

2. A sequence complementary to each of the code's triplets, which enables each transfer RNA to pair with the corresponding triplet of messenger RNA.

The pairing comes about in association with a complex constituent, the ribosome, as it were the "workbench" upon which the various components of the mechanism are put together. The messenger RNA is read sequentially, an as yet imperfectly understood mechanism permitting the ribosome to move, triplet by triplet, along the polynucleotide chain. In its turn each triplet pairs on the surface of the ribosome with the corresponding messenger RNA carrying the amino acid specified by that triplet. At each stage an enzyme catalyzes the formation of a peptide bond between the RNA-borne amino acid and the preceding amino acid at the end of the already formed polypeptide chain, thus lengthened by one unit. After which the ribosome moves one triplet further and the process is repeated.

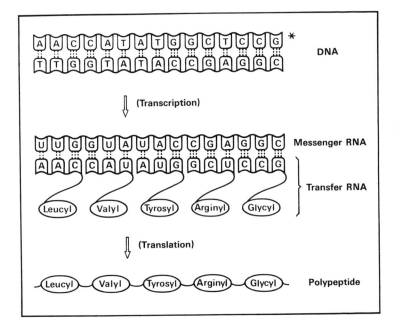

The figure above outlines the mechanism whereby the information corresponding to an (arbitrarily chosen) DNA sequence is transferred. Here the messenger RNA is assumed to be transcribed from the DNA strand marked by an asterisk. In actual practice the transfer RNAs pair one after another with the messenger; for the sake of clarity, they are shown in this figure as all pairing simultaneously.

4
Note Concerning
the Second Law
of Thermodynamics

SO MUCH has been written on the meaning of the second law, on entropy, on the "equivalence" between negative entropy and information, that one hesitates to review this subject in a few brief paragraphs. One or two general remarks may however prove of use to some readers.

In the form originally put forward (by Clausius in 1850, as a generalization of Carnot's principle), the second law specifies that *within an energetically isolated enclosure* all differences of temperature must tend to even out *spontaneously*. Or again — and it comes to the same thing — within such a space, if the temperature is *uniform* to begin with, no differences of thermal potential can possibly appear in different areas of the whole. Whence the necessity to expend energy in order to cool a refrigerator, for example.

Now within an insulated and enclosed space at uniform temperature, where no difference of potential remains, no (macroscopic) phenomenon can occur. The system is *inert*. In this sense we say that the second law specifies the inevitable *degradation* of energy within an isolated system, such as the universe. "Entropy" is the thermodynamic quantity which measures the extent to which a system's energy is thus degraded. Consequently, according to the second law every phenomenon, whatever it may be, is necessarily accompanied by an increase of entropy within the system where it occurs.

It was the development of the kinetic theory of matter (or statistical mechanics) that brought out the deeper and broader significance of the second law. The "degradation of energy" or the increase of entropy is a statistically predictable consequence of the random movements and collisions of molecules. Take for example two enclosed spaces at different temperatures put into communication with each other. The "hot" (i.e., fast) molecules and the "cold" (slow) molecules will, in the course of their movements, pass from one space into the other, thus eventually and inevitably nullifying the temperature difference between the two enclosures. From this example one sees that the increase of entropy in such a system is linked to an increase of *disorder*: the fast and the slow molecules, at first separate, are now intermingled, and the total energy of the system will distribute statistically among them all as a result of their collisions; what is more, the two enclosures, at first discernibly different (in temperature) now become equivalent. Before the mixing, work could be accomplished by the system, since it involved a difference of potential between the enclosures. Once statistical equilibrium is achieved within the system, no further macroscopic phenomenon can occur there.

If increased entropy in a system spells out a commensurate increase of *disorder* within it, an increase of order corresponds to a diminution of entropy or, as it is sometimes phrased, a heightening of negative entropy (or "negentropy"). However, the degree of order in a system is definable (under certain conditions) in another language: that of information. The order of a system, in such terms, is equal to the quantity of information required for the *description* of that system. Whence the idea, propounded by Szilard and Léon Brillouin, of a certain equivalence between "information" and "negentropy" (see pp. 59–60). An exceedingly fertile idea; but which may give rise to ambiguous generalizations or assimilations. Nevertheless it is legitimate to regard one of the fundamental statements of information theory, namely that the transmission of a message is necessarily accompanied by a certain dissipation of the information it contains, as the theoretical equivalent of the second law of thermodynamics.

A Note About the Author

Jacques Monod, together with André Lwoff and François Jacob, was awarded the Nobel Prize for Medi-
cine and Physiology in 1965 for elucidating the replication
mechanism of genetic material and the manner in which
cells synthesize protein.

Dr. Monod is now the director of the Pasteur Institute
in Paris, whose Cellular Biochemistry Service he created
in 1954 and directed thereafter. He was appointed a
professor at the College of France in 1967, and he is a
foreign member of both the Royal Society and the National Academy of Sciences.

A Note About the Translator

Austryn Wainhouse, an American who lives in France,
has also translated works of Georges Bataille, Simone de
Beauvoir, Jean Cocteau, and the Marquis de Sade.

A Note on the Type

The text of this book was set on the Linofilm in a face called Times Roman, designed by Stanley Morison for The Times *(London) and first introduced by that newspaper in 1932.*

Among typographers and designers of the twentieth century, Stanley Morison has been a strong forming influence, as typographical adviser to the English Monotype Corporation, as a director of two distinguished English publishing houses, and as a writer of sensibility, erudition, and keen practical sense.

This book was composed by the Poole Clarinda Company, Chicago, Illinois, and printed and bound by The Book Press, Brattleboro, Vermont.

Typography and binding design by Constance T. Doyle.